Project Management
in Construction

Project Management in Construction

Fifth Edition

Anthony Walker
BBS, MSc, PhD, FRICS

Emeritus Professor of Real Estate and Construction
University of Hong Kong

Blackwell
Publishing

Editorial Offices:
Blackwell Publishing Ltd, 9600 Garsington Road, Oxford OX4 2DQ, UK
   Tel: +44 (0)1865 776868
Blackwell Publishing Inc., 350 Main Street, Malden, MA 02148-5020, USA
   Tel: +1 781 388 8250
Blackwell Publishing Asia Pty Ltd, 550 Swanston Street, Carlton, Victoria 3053, Australia
   Tel: +61 (0)3 8359 1011

First edition published by Granada Publishing 1984
Second edition published by BPS Professional Books 1989
Third edition published by Blackwell Science 1996
Reprinted 1997, 1998, 1999
Fourth edition published by Blackwell Science 2002
Fifth edition published by Blackwell Publishing Ltd 2007

ISBN: 978-1-4051-5824-4

Library of Congress Cataloging-in-Publication Data

Walker, Anthony, 1939-
   Project management in construction / Anthony Walker. – 5th ed.
      p. cm.
   Includes bibliographical references and index.
   ISBN-13: 978-1-4051-5824-4 (pbk. : alk. paper)
   1. Construction industry–Management. 2. Project management. I. Title.

   HD9715.A2W27 2007
   624.068′4–dc22
                                                            2007006952

A catalogue record for this title is available from the British Library

Set in 10/12 pt Sabon
by Aptara Inc., New Delhi, India
Printed and bound in Singapore
by Utopia Press Pte Ltd

The publisher's policy is to use permanent paper from mills that operate a sustainable forestry policy,
and which has been manufactured from pulp processed using acid-free and elementary chlorine-free
practices. Furthermore, the publisher ensures that the text paper and cover board used have met
acceptable environmental accreditation standards.

For further information on Blackwell Publishing, visit our website:
www.blackwellpublishing.com

Contents

Preface

The management of construction projects continues to be a challenging occupation as the environment in which projects are constructed increases in complexity. Such complexity arises from the increasing technological, social and economic forces that impact on the process of constructing the world's buildings and infrastructure. The fifth edition of this book addresses this increasing complexity by examining the advances that have taken place in the major theoretical concepts which underpin the discipline of project management and by considering practical developments in the context of the underlying theory. Thereby the book retains the focus on project management theory of previous editions, in contrast to most other project management texts, which continue to be mainly concerned with procedures and techniques.

The original components of project management theory developed in previous editions – essentially systems theory and transaction cost economics set within the context of organisation theory – remain relevant. They are enhanced by updating and are complemented by an examination of the relevance of new ideas such as chaos and complexity theory to construction project management. Certain aspects of organisation theory meriting further attention, such as trust and hierarchy, have been further developed. The updating of the text relating to organisation theory generally has once again relied to a great degree on Scott's book, this time the 2003 edition.

The issue of sustainability is introduced, essentially as an example of a powerful environmental force impacting on the construction industry itself, the project management process, and – importantly – upon clients, and hence being transmitted to the process. The presence of a 'new' environmental force of such strength requires adaptation by the systems upon which it impacts, yet the construction industry has hardly started to react.

Major 'advances' in the practical aspects of construction project management are examined in the light of theory to establish whether they can justifiably claim to improve the project management process. 'Advances' examined in this way include supply chain management, clusters, public–private partnership/private finance initiatives (PPP/PFI), prime contracting and programme management. Also, recognition of the place of stakeholders in the project management process is made more explicit whilst still maintaining the focus of project management in this book on the perspective of the 'conventional' client.

The opportunity has been taken to reorganise the internal structure of some chapters and two chapters have been merged to improve coherence.

The longevity of this book appears to show that both practitioners and students have found it useful, and for that I am grateful. Its purpose is to give a depth of understanding to the construction project management process by showing that within organisation theory there are powerful theoretical bases which, when applied with discrimination to the construction management process, can give real insights that cannot be achieved by considering solely the practical manifestations of the process.

Once again I wish to put on record my indebtedness to my colleagues at the University of Hong Kong, particularly Chau Kwong Wing, Steve Rowlinson, Anita Liu and Richard Fellows, for their invaluable advice and assistance over many years. For the fifth time my thanks are due to my wife for her increasing tolerance and encouragement. I feel I must also reiterate my gratitude to Alan Wilson and Arthur Britch, who started me down this road those many years ago as well as to the many colleagues who publish in the field of management and economics of construction and upon whom I and many others rely for our knowledge. Nevertheless I remain solely responsible for any faults that remain in this book, but hope that it continues to make a contribution to the field.

Anthony Walker
Hong Kong

Chapter 1
Relevance of Organisation Structure

Introduction

The management of construction projects has been carried out since people first co-operated to erect buildings yet there is little documented knowledge of how people interact in this process. It is revealing that historical and contemporary accounts of construction work pay little attention to how people worked together and managed their activities.

Writers over the ages have concentrated upon the buildings themselves, particularly on aesthetics, the use of new materials, technological developments and the impact of buildings on their environment. How people were organised and managed received scant attention until recent times. What was written tended to be about such charismatic characters of enormous ability such as Brunel and Wren, and not about how they structured their organisations. Refreshing in this respect is Alan Wilson's (1976) account of how the construction of Rhuddlan Castle was managed by Master James of St. George. He completed the works using 3000 men in less than three years, and that was over 700 years ago. Perhaps we would see a significant improvement in performance if the managers of the construction process today were given titles like 'Master James'! A more contemporary contribution is made by Morris (1994) who documents how many more recent projects were organised and implemented. Many are outside the construction field and include, for example, Polaris and Concorde, but these illustrations of project management in practice also make a valuable contribution to our understanding of project management in construction.

The way in which the available skills are used is of paramount importance in providing what clients expect from their projects. There is little point in the construction industry developing the special skills of its members if no one is going to amalgamate them in the best manner to meet a particular client's objective.

The traditional method of organisation for construction projects, by which is meant one in which the architect or engineer is designer and manager of the process, using specialist consultants with the construction contract awarded by competitive tender after the design is substantially complete, evolved in response to pressures other than solely the needs of clients. It developed in conditions that were considerably more stable than those faced today by both the construction industry and its clients.

The complexity of the conditions within which the construction industry's clients exist makes them place increasing demands upon the industry in terms of the performance of projects (both functionally and aesthetically), the capital and running costs, environmental and sustainability demands and the time required from conception of the project to occupation. This has come about as a result of technological developments, globalisation, uncertain economic conditions, social pressures, political instability, etc. Such forces have led to the emergence of stakeholders in projects: that is, organisations, institutions and individuals that are not formally clients but can claim a socially acceptable interest in projects to which clients are required to acknowledge and respond.

Within such conditions, clients from both private and public sectors have to increase their effectiveness to remain competitive and to satisfy their own clients who transmit the demands of a complex world to them. The construction industry has in turn to respond to demands from clients that arise from such conditions and is itself also subject to external pressures in a manner similar to that of its clients. It therefore needs to respond by mobilising the talents it possesses in a way which recognises the particular needs of individual clients. It has become clearly recognised that it is unreasonable to suppose that the traditional way of organising construction projects is likely to be a universal solution to producing a project in today's conditions.

The complexity of clients' demands, together with the increasing complexity of building, civil and industrial engineering, and other constructional work, particularly as a result of technological developments, has over the years resulted in increasing specialisation within the construction industry. The professions associated with construction emerged as separate skills (e.g. architecture, quantity surveying, structural, mechanical and electrical engineering, acoustics, safety), as have the many specialist subcontractors working with the main contractor. On any project, even a small one, large numbers of contributors and skills are involved. On the largest there is a vast range of skills and materials required and an enormous variety of people and equipment to mobilise. Where these projects are carried out overseas, there are many additional problems of logistics and language.

The key to the management of construction projects is therefore the way in which the contributors are organised so that their skills are used in the right manner and at the right time for the maximum benefit to the client. There is little point in the construction industry developing its skills if they are not then implemented effectively.

The way in which the industry and its skills and professions evolved has compounded the problem of organising effectively. Specialisation has been accompanied by the creation of independent companies offering the specialisations, and the complexity of construction has led to greater interdependency between the specialisations and hence between companies. This produces a high level of differentiation within the construction process and a consequent need for strong integration of the independent companies and skills. This situation was reinforced by professional allegiances which, in the UK and elsewhere, were compounded by the establishment of professional institutions, which in turn contributed to the division of the design professions among themselves and their separation from construction firms.

It is against this background that the traditional solution to project organisation has attempted to cope with increasing complexity and uncertainty. The strain has shown through the development and increasing use of alternative approaches such as design-and-build, management contracting and construction management and initiatives such as partnering and prime contracting. There are now many alternative forms of organisation for construction projects, but what is needed is a framework for designing organisations to suit the particular project and the conditions in which it has to be executed. Pressure from clients has made the professions and industry take more seriously the need for organisation design, which is fundamental to the improvement of the project management process.

It should be clear by now that this book views the most important element of project management as an organisational issue which incorporates the way in which people are organised and managed in the project management process. This is a long step from the view of project management still taken by many which sees it as a collection of planning and control techniques and other management tools and decision-making techniques. The distinction is important as the use of techniques and tools, however sophisticated, will be of no avail if they are applied within inappropriate organisations seeking to achieve misguided objectives. Objectives and organisation must come first if the use of planning and control techniques is to be effective in providing the information on which management decisions can be based.

Whilst the terminology in this book is drawn from building rather than civil engineering, the application of organisation theory is as relevant to civil engineering as it is to building. The design of both civil engineering and building project organisations will benefit from the application of the ideas arising from the issues discussed in this book. Project management has, over recent years, been taken much more seriously in both sectors demonstrating the parallel need identified by sponsors and managers of projects. Further progress will be made through a greater understanding of the basis of project management which will arise from the common view of theoretical work identified in this book even though the applications relate predominantly to building.

Management and organisation

Before discussing project management and particularly organisation structure, it is necessary to have a clear idea of what is meant by management and by organisation. It is hardly surprising that definitions of management have occupied authors of management literature at length when the *Shorter Oxford English Dictionary* lists ten meanings of 'to manage', ranging from 'training a horse', and 'wielding a weapon' to 'controlling the course of affairs by one's own action'. The minds of many are also conditioned by its ironical use, which the dictionary quotes as 'to be so unskilful or unlucky as to do something'. Much of the literature presupposes that the reader has a clear idea of the concepts of management and organisation. Some writers offer a dictionary-style definition, but the operational definitions offered by Cleland and King (1983) are perhaps still the most useful.

An operational definition is one that identifies a number of observable criteria, which, if satisfied, indicate that what is being defined exists. Cleland and King's operational definition of *management* identifies the criteria of 'organised activity, objectives, relationships among resources, working through others and decisions'. In providing an operational definition of *organisation*, Cleland and King had to employ many of the elements used in their management definition. Organisation and management are intrinsically interlinked concepts. The former is concerned with the 'organised activity' part of their definition of management, and their observable criteria are 'objectives, some pattern of authority and responsibility between the participants with some non-human elements involved'. Decisions, both routine and strategic, are required from management to make the organisation operate.

Although management and organisation are closely interlinked concepts, it is interesting to note that management is more frequently defined in the literature than organisation. Yet it was said (Likert 1961) many years ago that 'how best to organise the efforts of individuals to achieve desired objectives has been one of the world's most important, difficult and controversial problems' and it still holds today. It may be said that, in industries more homogeneous than the construction industry, the distinction between management and organisation is sufficient, but an especially sharp focus on the organisation of the many diverse contributors to construction projects is necessary if the successful management of projects is to be achieved.

For the purpose of accomplishing a construction project, an organisation can be said to be the pattern of interrelationships, authority and responsibility that is established between the contributors to achieve the construction client's objectives. Management is the dynamic input that makes the organisation work. When this takes place, the organisation ceases to be static and works and adapts to meet the objectives laid down for it. Management is therefore concerned with setting, monitoring and adapting as necessary the objectives of the project organisation as transmitted by the client (which includes stakeholders), and with making or advising on the decisions to be made in order to reach the client's objectives. This is achieved by working through the organisation set up for this purpose, which is particularly difficult for construction projects owing to the temporary nature of most project organisations. In many cases, members of the organisation are part-time, as they are also involved in other projects and are normally seconded from their parent company.

The contributors to the project act through the organisation that has been established to carry out their work, and they produce information that allows the managers of the project to make the decisions that will keep the process going. The effectiveness of the organisation structure is therefore fundamental to the quality both of the information on which decisions will be taken and of the decision-making process itself.

Definition of construction project management

General management definitions require amplifying before they can be used for defining construction project management, which can be said to be:

> 'The planning, co-ordination and control of a project from conception to completion (including commissioning) on behalf of a client requiring the identification of the client's objectives in terms of utility, function, quality, time and cost, and the establishment of relationships between resources, integrating, monitoring and controlling the contributors to the project and their output, and evaluating and selecting alternatives in pursuit of the client's satisfaction with the project outcome.'

In this context, resources is a general term, which includes materials, equipment, funds and, in particular, people. A fault with many definitions of project management is that they do not make a specific reference to managing people to achieve a project. Although it can be implied that projects can only be achieved by working through others, nevertheless it is important that definitions make explicit reference to this fundamental aspect of project management.

The implementation of this definition could take many forms in practice, depending on the nature of the project and the circumstances in which it is carried out. However, as referred to earlier, the professions still tend to seek to achieve it through a traditional organisation structure even though many new structures have been devised. Nevertheless, no matter what organisation structure is adopted, if project management is taking place, the activities identified within the definition should be observable. The elements of project management have not changed fundamentally since this definition was devised in the early 1980s as illustrated by the definitions used by the British Standards Institution (2000) and the Chartered Institute of Building (2002).

In this context, it is important that the objectives of project management are not drawn too narrowly in terms of only time and cost. Interestingly, Munns and Bjeirmi (1996) attempt to draw a distinction between the project on one hand and project management on the other. They argue that project success is about client satisfaction with the longer-term performance of the project fulfilling its purpose, and not short-term time and cost objectives that they believe are often perceived as the objectives of project management. This distinction takes a narrow view of project management. The objectives must include client satisfaction, which should incorporate the client's view of its need to satisfy stakeholders, in all aspects of project success that encompass the function and utility of the project, which should be broadly defined and long term.

Objectives and decisions

Reference to objectives and decisions has been predominant so far and they have particular significance for construction project management. The objectives of the project management process are those of the client, and a role of project management is to ensure that the project organisation works to achieve the client's objectives. Similarly, decisions taken during the process should be taken with the sole purpose of achieving the client's objectives.

As referred to earlier, the term 'stakeholders' has been used over recent years as a supplement or alternative to 'client' to demonstrate that there can be many parties with a formal or informal interest in a project. Whilst this is recognised, generally this book will continue to use the term 'client' to indicate the organisation (of which there may be more than one) with the authority to take decisions that the project team must follow.

Because a large number of organisationally independent firms are usually involved in construction projects and second their staff on a part-time basis, their integration and orientation to the client's objectives are major functions of project management. Thus, objectives need to be clearly stated and the head of the project management team will have to extract them from the client, state them clearly and transmit them equally clearly to the contributors to the project.

It is important that any adaption of the objectives that may subsequently occur is treated similarly. It is natural to be greatly concerned with the original objectives, but adaptions are not always given the same attention, leading to dissatisfaction with the completed project on the part of the client.

As contributors will normally be involved in a number of projects at the same time, conflicting demands upon their time and attention are always likely to occur within contributors' firms. The project organisation must be designed and managed to resolve such conflict in the interest of its client so that it does not detract from the achievement of the client's objectives.

The project management process and the project manager

The use of the title 'project manager' in the construction industry has deflected attention away from consideration of the process of project management. It is necessary at this point to distinguish between the title and the process. A common reaction seems to be that if there is someone called a project manager, then all project management problems will instantly be solved. But the project management process will take place irrespective of the titles of the people in the process. The industry needs to be concerned with identifying and studying the process of managing construction projects and with structuring its organisations and implementing techniques and procedures that make the process more effective. It may well be that the designation of a suitable individual with the title of project manager will assist in this, but it is not likely to be an instant and universal solution.

The approach should be to identify the process to be undertaken for the achievement of the specific objectives of the client, the conditions in which it is to be carried out and the people available for the project. As a result of this analysis, the organisation structure should be designed to suit the particular project. The nature of the project should establish the roles of the contributors and ascertain whether or not a role emerges that requires the title of 'project manager' as a reflection of the project's needs. Such an approach would focus attention upon the process of project management with the result that effort should be put into making it more effective rather than into a preoccupation with titles.

The title 'project manager' should have a reserved meaning in the construction industry. Projects are executed for clients and as the title means managing the project as a whole, then it should refer to managing the project for the client: that is, the specific and unwavering objective of the project manager must be achievement of the client's objectives. The project manager will therefore seek to resolve conflict in the process in the interests of the client. This implies that ideally the project manager should be a member of the client's organisation.

One step removed from this, and more practically, the project manager could be acting as a professional consultant without an entrepreneurial interest in the project. Even in this latter case it is possible to conceive of a situation in which project managers might have difficulty in resolving conflict solely for the benefit of the client if, for instance, they are handling a number of projects that generate conflicting demands on their time and attention. Any further removal of the project manager from direct responsibility to the client makes the title difficult if not impossible to justify.

The title does not always have this reserved meaning in practice and this leads to confusion. Other titles are available which can be used to imply the orientation of the particular management activities undertaken. For instance, construction manager, contract manager, design manager are roles that are often designated project manager. The activities implied by such titles do not necessarily have the client's interest as their main concern. It must be added that they do not, of course, deny satisfying client objectives as one of their objectives. The increasing use of design-and-build structures confuses the issue somewhat as a design-and-build company will usually designate its own person as project manager but this role is distinctly different from the client's project manager (either in-house or in a consultancy role) as its focus is not solely on the client's objectives.

To complete the array of management activities in the construction industry, it should be recognised that general management of the contributing firms will also be taking place, the objective of such activities being the effectiveness of the firm.

Projects, firms and clients

Conflicting objectives

The work of firms in the construction industry and its professions present two types of management issue: the problem of managing firms and that of managing projects. This leads to a rather complex matrix management structure, shown in a much simplified form for a traditional arrangement of contributors in Fig. 1.1. This diagram is greatly simplified because it implies that the three projects are each being undertaken by the same professional practices, general contractor

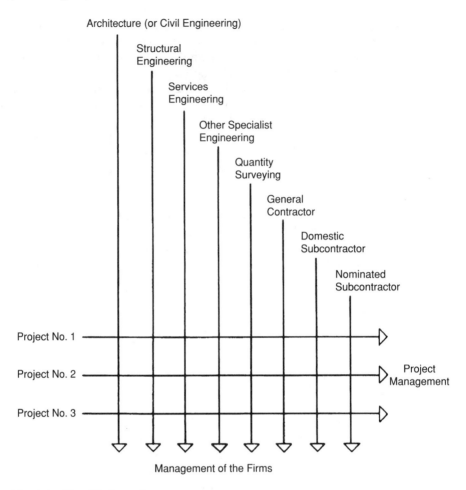

Fig. 1.1   Simplified matrix management structure.

and subcontractors. In practice, of course, this is rarely the case. Normally there will be different mixes of professional practices, general contractors and subcontractors on each project. Even if the private practices are the same, by using competitive tendering it is very unlikely that the general contractor and subcontractors will remain the same. Such a lack of consistency of contributors makes it extremely difficult to improve the effectiveness of the project management process. Not only do firms have to get used to each other at both a corporate and individual level, but they are also unlikely to invest much time and money in making the process more effective when they know that any temporary management structure that they establish may only occasionally be used in a similar form again. Relational contracting initiatives have gone some way to overcoming such problems for clients with multiple projects, but the larger proportion of projects still takes the traditional form or something similar.

Construction projects, whether traditional or relational, are therefore usually undertaken by an amalgam of firms, which change from project to project. The firms involved in each project are independent companies, which are organisationally interdependent in terms of the project. This situation creates a potential for conflict between the needs of each firm and of each project. Each firm has objectives which are expressed in terms concerned with the efficiency of the firm, such as:

- increasing productivity
- improving service
- maintaining existing clients
- attracting new business.

The major purpose is to improve effectiveness and hence service and profits. Professional practices would claim to be less entrepreneurial than contracting organisations, but nevertheless conflicts between the needs of individual firms and the needs of projects will still arise. For instance, what does a firm do if there is a choice to be made because of limited resources between progressing an urgent matter for an existing client and undertaking a piece of work that could clinch a commission with a new client? Similarly, what would a contractor do if faced with a choice between keeping a piece of equipment on site to be used to keep a project on programme and removing it to another site in order to increase the profit on the second site, knowing that liquidated damages are unlikely to be claimed on the first site?

The objectives of project management, which ideally should also be the objectives of the firms involved in the project are, as has been said, the objectives of the client. These will relate directly to the project and will be:

- functional satisfaction
- aesthetic satisfaction
- completion on time
- completion within budget
- value for money.

Where, then, does the responsibility lie for ensuring that the project's objectives are met? The professional practices, particularly architects and civil engineers in a traditional arrangement, would say that it rests with them. But who, then, is to resolve any conflicts that may occur in a manner which is to the benefit of the particular project? If the practices are to do it, can they be sufficiently unbiased to resolve conflict to the benefit of the project to the extent to which the client may require?

The matrix structure using independent firms seems to need the responsibility for project management to rest in a firm or individual who is independent of the potential conflicts within the contributing firms. But may not such a firm also be faced with similar conflicts if dealing with a stream of projects?

Ideally it seems that project management should be exercised by the client organisation itself, and this reflects the need for clients to be very close to the organisation and implementation of their projects. However, many clients do not have the expertise to manage their own projects. This, therefore, is the dilemma for clients and for project management. Clients should be concerned to ensure that the design of organisation structures for their projects recognises and seeks to overcome such potential conflicts.

These issues raise the question as to whether the construction industry is unique. There have been a number of initiatives to transfer techniques from other industries, with mixed success; e.g. lean construction, prime contracting and requirements management (Fernie *et al.* 2003). As Fernie *et al.* argue in relation to requirement management, practices are embedded in context. The structural differences between industrial sectors mean that practices have limited meaning when transferred. However, the construction industry is not unique in terms of its organisational problems except perhaps that these pervade all levels of construction activity and in many countries are firmly rooted in the historic development of the industry and its professions. Organisation theory, being generic, as opposed to techniques, is relevant to analysing these issues.

Interestingly, recognition of the project management process as a suitable subject for formal study and research emerged from complex projects outside the construction industry (Morris 1994), for example in connection with the defence/aerospace programme in the USA and in other industries facing complex demands that require inputs from a range of independent firms, yet the situation they faced has been present in construction certainly since the nineteenth century.

It has taken the complexity and constraints of today's world, together with initiatives from outside the industry, to focus attention upon the way projects are managed as a possible means of finding solutions to some of the problems the industry faces. This reflects much of what Marian Bowley (1966) deduced about the inertia within the building industry and professions which stifled innovation.

Organisation patterns

Frequent reference has been made to the traditional arrangement of independent professional firms and contractors selected in competition. However, variations in organisation patterns have grown substantially in recent years through the growth of, for example, interdisciplinary group professional practices, design-and-build companies, and joint ventures for overseas contracts all of which seek to overcome some of the problems associated with the differentiation of professional firms contributing to projects. Similar advantages could also accrue to interdisciplinary public sector groups but, on the other hand, if a public service department provides only one type of professional service for their project (e.g. architecture) and uses professional firms for other services, the problems of organisation may be even more complex.

Different organisation patterns are also generated by the way in which the general contractor and subcontractors are selected. A range of alternatives is available, for instance, design-and-build, two-stage tendering and negotiated contracts and it is necessary to ask whether a proposed contract generates the organisational form most likely to achieve the client objectives. It has been found that it is unlikely that there is one 'best' form of contract for construction projects (Nahapiet & Nahapiet 1985). One then has to add to this what is probably the most significant variable: the vast range of client types served by the industry. Whilst not an organisation structure, partnering is a behavioural initiative aimed at developing a closer working relationship between client and contractor.

Variety of clients

Clients vary in many ways. Of particular importance is the variety of objectives that clients seek to satisfy. Differences in this respect are particularly marked between private and public sector clients, and overseas and multinational clients may have objectives rarely encountered in home markets.

The variety of objectives is compounded by the range of uncertainty of clients' objectives. The construction industry and its professions have to be skilled at translating such variability in a way which enables them to produce projects that satisfy their clients. They have to deal directly with their clients and in order to do this, and obtain and interpret instructions properly, they need to understand how their clients' organisations operate as the organisation structures used by clients vary considerably to reflect the needs of clients' major activities. As everyone, either individually or corporately, is a potential client for construction work, the construction industry and its professions could be called upon to work with every or any possible organisation configuration. The industry and its professions need to understand how organisations work in order to organise themselves and also understand how their clients' organisations work, so that they may be in the most advantageous position to interpret and implement their clients' objectives.

The demands that both private and public clients place upon the construction process are frequently complex and uncertain. This simply reflects the complexity and uncertainty of the modern world, as demonstrated by contemporary economic, social and environmental issues. The construction industry and its professions are themselves also subjected directly to such forces.

The professions and industry are now more readily developing approaches to the design of organisation structures that are tailored to satisfying specific client objectives, and take advantage of the range of temporary management structures available. What is needed is a framework for designing the most appropriate structure.

The contribution of organisation structure

The concept of organisation structures has been said by Hinings (2003) to be at the heart of organisational studies. He believes structure is important because it has been historically central to the development of organisation theory, continues to be of importance to existing and new organisational forms, and is a prime analytical construct for organisational theorists because it is central to the thinking of managers. There are many factors other than organisation structure that have a significant bearing upon the performance of an organisation. However, organisation structure is a particularly important aspect as, if properly designed, it allows the other aspects to function effectively.

This is not to say that, if an organisation is inappropriately designed, it will not perform adequately, as people have the ability to construct informal organisation structures that circumvent the formal structure often to the benefit of performance. However, a strong informal structure can work against organisation co-ordination and control. The ideal is when the organisation is sufficiently well designed that it does not generate an informal structure. Such an outcome would mean that the organisation is designed to meet its specific objectives and that the participating members would have confirmed that, in their view, this is in fact the case.

An appropriately designed organisation structure for a project will provide the framework within which the other factors that influence the effectiveness of the project management process have the best chance of maximum performance in the interests of achieving a client's objectives. For the purpose of construction project management, the major internal factors influencing the effectiveness of the management process can be considered to be:

- behaviour
- techniques and technology
- decision making
- organisation structure.

These aspects are interrelated and interdependent, as illustrated in Fig. 1.2.

The project management process is also subject to external influences. These comprise all elements outside the process which, if they change, demand a response from the project management process if it is to remain effective. Examples are economic forces, which may affect the client and modify his objectives for the project, and legal forces, which may require changes to the design, e.g. revised building regulations.

Behaviour

The behavioural factor consists of the characteristics of the individual members of the organisation as reflected in their motivation, reaction to status and role relationships and their personal goals and values. It therefore determines the attitude they have to their work on the project and to the work of others. Attitudes are significantly affected by external influences (e.g. the views of society) in addition to being influenced by the other aspects of the management process, for example Loosemore (1994) finds that high reciprocal interdependency of tasks frequently found on construction projects forces people to find solutions to problems. Behavioural responses have particular significance for construction project management because of the sentience of the various professions and skills involved, many of which have strong allegiances and view projects from very different positions, as illustrated in Fig. 1.3. It is a factor that can have a significant impact on the effectiveness of the project management process.

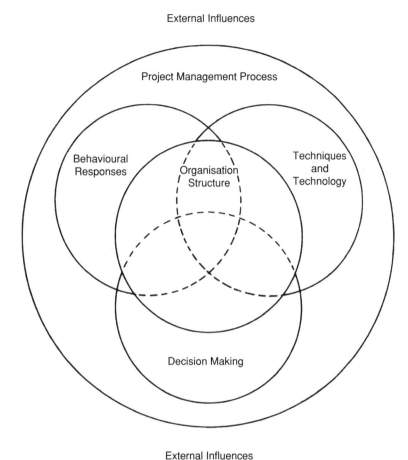

External Influences

Project Management Process

Behavioural
Responses

Organisation
Structure

Techniques
and
Technology

Decision Making

External Influences

Fig. 1.2   Factors in the project management process.

Techniques and technology

Techniques and technology are the tools used by the members of the organisation to produce the building or other construction work. The quality of the tools they use is determined by the knowledge the project team have of the techniques and technologies available and their skill in using them. The interdependency of the organisation structure and the techniques and technology used is based upon the need for the organisation to be structured in such a way that the appropriate techniques and technologies are drawn upon and used at the correct time in the process of designing and constructing. As a corollary, the techniques and technologies adopted may demand a certain organisation of the contributors to make their use effective. The techniques employed and the way in which they are put together by the project management process are fundamental to achieving clients' objectives. They encompass evaluation, appraisal and control methods,

Fig. 1.3   Perspectives of the contributors (original idea and sketches by Dave Taylor).

contractual techniques and approaches to design, as well as the techniques of constructing the project. Particularly important for construction projects is the project information facility employed and how this relates to the organisation structure.

Scott (1992) believes that three dimensions of technology – complexity, uncertainty and interdependency – are most important in explaining differences in structural characteristics of organisations. Technical complexity leads to structural complexity and increased professionalisation, uncertainty leads to lower formalisation and decentralisation of decision making and complexity and uncertainty to greater interdependency needing higher levels of co-ordination. From this analysis the links of technology to organisation structure and decision making are clear to see, particularly in construction.

Decision making

Although the quality of decision making is vital for all organisations, it acquires special status on construction projects. The complexity of projects is reflected by the large number of specialists who contribute to the decision-making process. It is therefore closely related to the organisation structure, which determines how people work together to produce the output that forms the basis on which decisions are made.

In moving towards the completed building, the process is characterised by a series of 'pinch points' through which it must pass if it is to make progress. At each of the points, a decision has to be made which could include the option to abort the project. The process of making these decisions will be managed by the project management process as a whole and will be based on output generated by the contributors working within the organisation structure using techniques and technologies. The client and managing executive will take many of these decisions based on the advice of the specialist contributors. The significance of decision making is that it should be interrelated with the organisation structure in such a way that advice is received by the decision maker from the appropriate contributors at the appropriate time.

Organisation structure

The organisation structure of a particular construction project is a subset of the project's management process. It structures the relationships of the members of the organisation and hence influences their responses to the demands placed on them. It establishes the way in which advice is generated for decision making and the use of techniques and technology in the process. It should be designed to allow these factors to be integrated.

The managing executive of the project should be responsible for designing the organisation structure and should provide the integrating activities that weld the parts into a unified whole. The managing executive then provides the dynamism required to make the whole process seek to achieve the client's objectives.

Organisation theory and project organisations

Organisation theory recognises that professional organisations are distinctive. As Scott (1992) states:

> 'Certainly the most elaborate and intricate organisational arrangements yet devised for coping with high orders of complexity and uncertainty in production systems are to be found in the *professional organisation*.'

However, he continues by focusing on what he terms autonomous professional organisations and heteronomous professional organisations. The former are those in which the professionals have responsibility for their own goals and the establishment and maintenance of their performance standards. Examples are design firms, law firms, medical staff in hospitals and academics in universities. The heteronomous types are when the professional staff are subordinated to administrative staff and have relatively little autonomy. Public agencies fall into this category, such as schools and welfare agencies.

Scott recognises that autonomous professionals use a number of organisational forms, one of which is the project team. He recognises this in hospitals and university research projects. He concludes that:

'Organisations that in one way or another utilise lateral relationships as legitimate avenues of information and influence flows constitute the new generation of organisational forms. As we have attempted to illustrate, a number of different lateral structural arrangements are in use – including project teams, matrix structures, organic or clan systems, and professional organisations. All move us away from unitary hierarchial arrangements, "beyond bureaucracy" or "from bureaucracy to adhocracy".'

What is significant here is that the types of structure which have been used as a matter of course in the construction field were seen as the 'new generation of organisational forms' by organisation theorists. In a later edition Scott (1998) states:

'The "new" organisation is more closely attuned and interdependent with its suppliers and buyers; it is less likely to have a large body of permanent full-time employees, and much of its structure is shaped and the logics controlling its operation are provided by experts and specialists imported from outside who have less stake in the welfare of their specific employer . . . newer models recognise the reality of interdependences and stress the importance of developing appropriate alliances and alignments (see Ancona *et al.* 1996)'.

Once again, organisation theory based on general organisation forms appears to be moving towards the forms both traditionally and more recently developed in the construction fields such as temporary project teams (traditionally) and partnering (more recently). This shows the need for an organisation theory related directly to the forms used for construction projects. Whilst elements of general organisation and management theory are relevant and useful to project management, that which is so needs to be carefully distilled and developed before it is able to provide insights which contribute to making project organisations more effective. The distinctions are highlighted by Thompson (1991):

'The temporary nature of the project team and the need to define and achieve specific project objectives against a demanding timescale, together with the high level of risk and expenditure encountered on many projects, will demand a style of project management that is likely to be more dynamic than that of corporate management.'

Against this backcloth, it may seem surprising that project organisations have been able to design and construct projects reasonably successfully for hundreds of years and particularly so in the more unstable environment of this century. But construct them they have; nevertheless this does not mean that they cannot do so more effectively. The emergence of a theory of project management is a necessary basis for increased effectiveness in which organisation theory will play a significant part. Such theory will emerge in part from the general management literature but also more particularly from experience in practice and from original thinking specifically about project organisation.

Project organisation in context

This book is primarily concerned with understanding the elements which lead to the design of effective organisation structures for construction projects. However, the systems approach to organisation design which is developed does not ignore the other major factors that influence the effectiveness of the project management process. Rather the systems approach provides the core to which all other elements relate. Such elements include the behavioural/psychological approaches to management particularly relating to motivation, organisation culture, strategic management, decision-making techniques and many other specific aspects of management together with transaction cost economics, all of which make significant contributions to an understanding of the management process and have many publications devoted to them.

Chapter 2
Evolution of Project Organisation

Introduction

The way in which construction projects are organised in different countries has evolved from traditions and conventions laid down in each country many years ago. The traditions and conventions of the UK have had a particularly wide significance as they have been exported to many parts of the world over the last two centuries. A brief account of project organisation evolution in the UK may help to explain the position reached in trying to develop more effective ways of managing construction projects. It will have been paralleled in many other countries.

Origins

The way in which contributors to construction projects organise themselves in the UK has its origins in the Middle Ages, but over the centuries the original patterns have been influenced significantly by the increasing complexity of the conditions in which building has taken place and by the construction industry's attempts to cope with the prevailing conditions.

The surviving records of building in the Middle Ages are for prestigious structures and mainly show that a master mason was responsible for acquiring and organising labour and material and for the technicalities of construction on the basis of an outline from the client. Alongside master craftsmen there often existed clients' representatives, many of whom did not have practical experience of building but who were among the few people who were literate and numerate. They were expert administrators and went under a variety of titles, such as surveyor, clerk of works and sacrist. The client would pay directly for the labour and material consumed.

This direct method predominated until towards the end of the seventeenth century, although there is evidence as early as the beginning of the sixteenth century of work being let on a contract or 'bargain' basis. This basic pattern probably had many variants as the recorded titles of people are confusing and the relative responsibilities difficult to determine.

The eminence of master masons led to the most eminent being appointed King's Mason with responsibility for oversight of the king's palaces and castles. They also acted as advisors on a number of projects, in a role akin to that of architect in later years.

The relatively stable conditions in which the 'building industry' existed in the medieval period did not create conditions for change in the organisational pattern of building work until demand for building began to rise in the sixteenth century, when the distinctive role of the architect began to emerge and more work began to be awarded on a contract or 'bargain' basis. Engineers were more concerned with mechanical devices for military purposes than with buildings, but through their work on fortifications and castles their influence on buildings began to be felt although master craftsmen developed their own empirical engineering and jealously guarded this knowledge.

The period from the sixteenth century to the Industrial Revolution saw many changes, which had profound effects on the organisation of building projects. England had become a principal trading country of the world and travel had awakened interest in the buildings of ancient Greece and Rome, leading to a demand for such designs. This led to the clearer identification of the role of the architect, and the associated complexity resulted in an increasing tendency to let building work on a contract basis, although 'architects' also often acted as developers. Further impetus to change was created by the Great Fire of London, which led to the 'measure and value' method of settling payments and the employment of separate measurers. There still appears to have been little application of formal engineering to building, although some road and bridge building took place.

The great surge in the demand placed upon the construction industry was generated by the Industrial Revolution. The accompanying prosperity created demands for housing, to accommodate both workers and owners, and for buildings for the new industries. The demand for improved transportation led to the development of new engineering and building techniques, and to further industrialisation and demand for buildings. In response to such demands, new materials were being developed, which allowed new building techniques to be devised.

These activities created a concentration upon the specialist skills of the members of the building industry. The importance of the engineer emerged; there was the further separation of the architect and builder as specialists; quantity surveying skills were more firmly identified; and engineering was subdivided into civil, mechanical and electrical skills. However, this was an incremental process and specialists often acted in dual capacities.

The new complexity of the conditions within which construction work was executed, with greater emphasis on economy, value and prestige, the complexity of new building materials and technologies and the developing skills of the building industry specialists themselves, created the need for greater specialisation among them. These pressures led to the establishment of societies for the discussion of common problems. Architectural clubs were formed in 1791, but clubs for civil engineers had been set up as early as 1771. In 1834 clubs were established for surveyors and for builders. Subsequently, to protect themselves from economic pressures on the one hand and from the unscrupulous on the other, the clubs developed, in the nineteenth century, into professional institutions as the means of defining their position and creating their public image through the acquisition of royal patronage. This further emphasised the separation of the skills associated with construction and so reinforced allegiance to specialist skills rather than the industry as a whole, and created the basis from which today's traditional organisational structure for construction projects had grown.

The period from the late nineteenth century to the First World War saw a continuing rapid increase in the growth of the building industry. This was accompanied by the rise of the general builder for both speculative and contract work, and the parallel emergence of specialist craft firms. This occurred in response to the need for organising ability and financial strengths required for the process of urbanisation and industrialisation.

The architectural profession was moulded on the social and aesthetic pattern of the eighteenth century, when architecture was considered one of the arts with the actual construction of the building being secondary. By the late nineteenth century, the idea that there should be any connection between architects and the mass of industrial buildings and working class housing seems to have been generally disregarded. Architects were, by then, concerned primarily with prestigious buildings. These attitudes were reflected in the 1887 supplementary charter of the Royal Institute of British Architects, which laid down that no member of the Institute could hold a profit-making position in the building industry and retain his membership.

This separation of architects and builders was accompanied by further separation of architects and engineers. The development of industrialisation and the position adopted by architects decreed that industrial building was the province of engineers but, at the same time, engineers were commonly employed to advise on the structure of architect-designed buildings in addition to their core work on infrastructure projects. Hence, architects were technically dependent upon engineers, but engineers were not dependent upon architects. Significantly, engineers did not exclude themselves from being principals of engineering or building firms. Further separation occurred when, in 1907, the Royal Institution of Chartered Surveyors instituted the Contractors' Rule, which prohibited its members from being employed by construction firms.

Bowley (1966) describes the pattern that emerged as 'the system' and believes that it had acquired a strong flavour of social class distinctions, architects being the elite. Engineers were associated with trade and industry, surveyors were on the next rung of the social hierarchy and builders were regarded as being 'in trade'. She believes that, as a result, aesthetic and technical innovations in the late nineteenth and early twentieth centuries were completely out of step with each other, which inhibited the development of the major technical innovations of steel-framed and reinforced concrete structures vis-à-vis other countries and created a conservatism in the construction professions.

Building activity between the First and Second World Wars was much greater than before 1914. However, there were no important changes in the way in which the design and production of buildings were organised, although the efficiency of site operations was enhanced, particularly through mechanisation. The period was one of consolidation of the main professions through the establishment of professional qualifications tested by examination and of codes of conduct, which raised their status and reinforced adherence to the established pattern of project organisation.

The lack of innovation in building in Britain in this period was brought about primarily as a result of the lack of a built-in mechanism in the organisation of design and construction that could create the necessary stimulus. The innovations that did take place tended to be outside the industry, particularly in the organisation of the building materials industry and in the materials themselves. In addition, there was great concern with housing needs and the switch from commercial speculative development to public development reinforced the prevailing pattern of organisation as this work also used the same structures. The pattern of organisation of design and construction does not appear to have been fundamentally questioned during this period, as reflected in the list of official government publications, none of which were concerned with organisation but which concentrated mainly on materials and housing.

Even present-day organisation arrangements for building projects and attitudes to innovations within the industry still reflect, to a degree, the conservatism generated by patterns laid down before the Second World War. However, following a succession of official reports on these topics, the professions and industry have increasingly responded to the demands of environments far more complex than that in which the patterns were originally established. The dramatic developments in transportation, communications, health care, manufacturing technologies and the associated economic, social and technological order have been important forces for client-led change in the construction industry.

A perspective of contemporary influences

The Second World War and post-war activity

The impetus to innovation provided by the Second World War was dramatic and focused upon the need for economy in labour and reduction in the use of materials in short supply. This need was demonstrated by the rapid adoption of prestressed concrete, prefabricated buildings and the tendency to replace steel with reinforced concrete. Wartime also generated the first governmental enquiry (HMSO 1944) directly concerned with the organisation of building work, which was the forerunner of reports that questioned the suitability and efficiency of the prevailing organisation of the construction process, but which focused on building rather than civil engineering. Nevertheless, this report accepted the established patterns and concerned itself, primarily, with tendering methods and arrangements for subcontractors.

Following the Second World War, the demands placed upon the building industry rapidly increased in complexity. The pressures created by the need for rebuilding in the aftermath of war were followed by an acceleration in complexity of demand through the development of the Welfare State, which required new and more advanced buildings. Also, the increased sophistication of industry required increasingly sophisticated buildings and there arose the need to redevelop cities to cope with a more technological age. One of the driving forces behind these demands was the increase in the relative importance of government-sponsored buildings and the consequently greater involvement of government in building. The tendency to build larger production units, arising from the development of large-scale organisations in industry, was another important force.

In spite of the substantial changes in demand placed upon the industry, the pattern of organisation of projects remained largely unaltered. Increased government sponsorship of building projects served to reinforce allegiance to the traditional pattern by the need for public accountability, which was seen to be satisfied by competitive tendering on finished designs. Nevertheless, there were some innovations in organisation patterns through the use of negotiated tenders and 'design-and-build', but the resistance to change of the established pattern is illustrated by the reluctance of public authorities to adopt selective, as opposed to open, tendering even though this had been strongly recommended in the Simon Report (HMSO 1944) and again in the Phillips Report (HMSO 1950). Other developments were concerned with improving the effectiveness of site operations, particularly through prefabrication and in-house construction.

However, the need for greater co-operation began to be recognised following the Phillips Report, which commented upon the ease with which variations could be introduced during construction, the problems created by drawings issued late, the extensive use of nominated subcontractors and the desirability of establishing a common basic education for all those involved in the design of buildings and their production.

Increasingly, discussion centred upon the need for greater co-operation between all parties to the construction process. This was further stimulated by the greater need for engineers to be involved in the more complicated buildings being demanded, the necessity for reliable cost control and an increase in the number of large building firms.

The difficulties of the traditional pattern of organisation in coping with the demands of modern construction, which were evident between the wars, were greatly intensified after the Second World War, but the greater spirit of co-operation within the industry that had begun to emerge took place against the backcloth of the existing traditions and was not concerned with a fundamental reappraisal of the structure that had been established. This situation was reflected in the next major official enquiry, the Emmerson Report in 1962, (HMSO 1962) which reiterated the findings of the previous two post-war reports regarding the need to improve co-ordination of the members of the building team.

The significant reports of the 1960s

Although it was concerned with supply and demand in the construction industry, standards of training, research and technical information, the Emmerson Report is particularly significant for its observations on relationships within the construction professions and industry, and with clients, and in connection with the placing and management of contracts. It identified a common criticism of the construction process as the lack of liaison between architects and the other professions and contractors, and between them and clients. It commented, 'In no other important industry is the responsibility for design so far removed from the responsibility for production.' The report pointed out that although a common course of initial study for designers and producers of buildings had been recommended in 1950, no practical steps had been taken by 1962. Emmerson came to the conclusion that there was still a general failure to adopt enlightened methods of tendering in spite of the recommendations of earlier reports. His recommendations in this respect led directly to the establishment of the Banwell Committee in 1962, to consider these issues in more detail.

Of the official reports, the Banwell Report (HMSO 1964), and its review Action on the Banwell Report (HMSO 1967) had a significant impact upon the construction industry and its professions. A particular concern of the report was the unnecessarily restricted and inefficient practices of the professions, leading to over-compartmentalisation and the failure of the industry and its professions to think and act together. The 1967 review found some progress on preplanning projects but noted that the professions had done little to 'de-restrict' their practices. The review was encouraged by the increase in selective tendering and urged further consideration of serial and negotiated tendering. The Banwell Report also related to civil engineering as well as building.

The Emmerson and Banwell Reports brought into sharp focus the need to reform the approach to the organisation of construction projects. They were accompanied by other reports making similar points. At the time, construction project management was seen to be a passive procedural activity but the movement towards a more dynamic integrated approach was being suggested by Higgins and Jessop (1965) in a pilot study sponsored by the National Joint Consultative Committee of Architects, Quantity Surveyors and Builders. They clearly identified that the problems of communication in the building industry were created to a large extent by attitudes and perceptions about the values of contributors to the building process. They were probably the first to suggest that overall co-ordination of design and construction should be exercised by a single person (or group). Concurrently, a review of the construction industry by the National Economic Development Council (1964) was calling for improvement in the management of the construction process, and the co-ordination of activities of the members of the construction team and the administrative framework within which they were working. A rather rhetorical report by the Institute of Economic Affairs (Knox & Hennesey 1966) was also condemning the restrictive practices of the professions.

This spate of activity and concern with the performance and organisation of the industry and its professions marked the beginning of a self-examination. It was induced, to a large degree, by external pressures that reflected the greater complexity of the influences at work upon the industry and its clients. The economic expansion of the early 1960s and rapidly developing technology and changing social attitudes were manifested in demands for more complex and sophisticated projects and a more economic utilisation of resources. These forces were transmitted to the industry through its clients and also directly affected its techniques and attitudes, but such self-examination was likely to be slow when undertaken in the presence of the polarisation of skills and attitudes inherent in the professional structure that had emerged over the preceding century.

The project manager and other organisation initiatives

During the 1960s and subsequently, progress was made in developing collaborative work and skills, and in instituting procedures that provided a variety of organisational patterns, particularly in connection with the introduction of the contractor at various stages of the design process. However, there was still a need in official reports in the 1970s (National Economic Development Office, 1975, 1976, 1978) to stress that more attention should be paid to structuring and managing project organisations to create conditions for co-operation between contributors. Each of these reports recognised the distinctive nature of the project management process and the role of the project manager, and reflected the changes in attitudes and views expressed since the mid-1960s. They arose from the distillation of the professions' and industry's experience of working with novel forms of organisation. The 1976 report recognised the need for further study which would analyse existing patterns in the use of alternative methods of organisation of the design and construction process.

The external pressures that caused the professions to reconsider organisational arrangements for projects were accompanied by challenges from the Monopolies and Mergers Commission in relation to their codes of conduct and fee scales. Further pressure was brought to bear through the definition and development of project management concepts and applications in other industries and the recognition by project management theorists that the concepts and techniques are applicable to construction. The professions' and industry's response to these influences reflected the manner in which the traditional structures emerged. Each sector pursued its own approach to project management while recognising rather reluctantly that the role of project manager was not the right of any one profession.

A reflection of the unco-ordinated empirical evolution of project management as an activity separated from design skills is given by the number of definitions that emerged. The Chartered Institute of Building's (1979) paper identified thirteen definitions. It commented that the confusion of terminology and usage was unsatisfactory, and proposed a further definition! It was, perhaps, to be expected that those writing on such an important emerging idea, which was contrary to their traditional backgrounds, should seek to express their ideas in their own words. However, this resulted in a range of definitions that tend to reflect the particular background and experience of the writer rather than a generalised definition of the concept. It is encouraging, as noted in Chapter 1, that definitions of project management have now achieved a good measure of consistency.

The empirical nature of publications on project management was reflected in their emphasis on defining the jobs to be done by a project manager at various stages of a particular project rather than identifying the concept and process of project management. Nevertheless, such publications have been useful in emphasising the patterns that can be adopted with advantage to the client.

Against this background of pressure for change in organisational approaches, a number of project-based initiatives emerged. The project manager idea was only one such idea which was used to cover a range of organisational patterns. Others included *management contracting, design and construct contracts* and *construction management* all of which seek to increase integration (particularly of the contractor) and which may or may not incorporate a project manager but which do not necessarily overcome the polarisation of professional attitudes.

The 1980s saw some recognition by the professions and industry of the forces working upon them. There was a shift from the government-sponsored reports of the 1960s and 1970s to initiatives from the private sector reflecting the shift in the political climate as a more pragmatic position was adopted. Clients were increasingly making their voices felt. Carpenter (1981) was typical of clients, stressing that the industry frequently adopted inappropriate organisation structures and the British Property Federation (1983) came up with its own system to impose on an industry which was not changing itself sufficiently quickly. Government reflected this pragmatism with so called 'client guides' to procurement (Department of Industry 1982; National Economic Development Office 1985) and practical comparisons of different approaches to development with the emphasis on speed of construction (National Economic Development Office 1983, 1987). Alongside these developments, the original edition of this book in 1984 and other texts on construction project management (Bennett 1985) were stressing the need for organisation structure to be tailored to suit the particular project.

Within this climate the UK professions and industry began to implement more flexible approaches to the management of projects. As a result, the 1980s saw more adaption to suit the needs of clients and a more ready acceptance of alternative arrangements than in any other decade. Nevertheless, Mohsini and Davidson (1992) estimate that 80% of all building projects in the USA were still procured by traditional processes and the figure was probably similar in the UK although Bresnen and Haslam's (1991) work implies that it may be somewhat less at about 70%.

Following the spate of reports during the 1970s and 1980s the process drew breath until the Latham Report (Latham 1994) which reinforced the pragmatic tone of the 1980s rather than the more philosophical tenor of earlier reports. Whilst its focus was predominantly on contractual matters and their impact on conflict, payments and cash flow, it nevertheless found room for an important section on project management. The changes which had taken place since the 1970s were clearly recognised by the report's comment that 'there is increasing (if sometimes reluctant) acceptance that Project Management, and a separate discipline of Project Manager, are permanent and growing features of the construction scene'. The report continued by recognising that the manner by which project management can be provided takes a great many forms which may or may not require someone with the title of project manager. Recommendations were made requiring a clearer definition of the role and duties of project managers.

Nevertheless, comments on project management are rather overshadowed by the contractual matters and also by the high profile target of 30% real cost reduction of construction costs by the year 2000. However, a focus of the working group set up to implement the cost reduction initiative was on organisational matters to answer such questions as: How do you decide what to build? What is the best way to set up a team? Whilst much progress has been made in adopting flexibility in structuring and managing projects it appears that some fundamental questions still require attention.

The Latham Report began its executive summary by stating that previous reports on the construction industry had either been implemented incompletely, or the problems had persisted, and many of Latham's recommendations reflected this as they were not new, although suggestions for achieving them may have been. A recommendation which was new and important to effective project management was that the use of Co-ordinated Project Information should be a contractual requirement. By contrast the report again recognised the need for common professional education in just the way that the Emmerson Report of 1962 did which was itself reflecting Phillips' Report of 1950. The progress, over 50 years on, has been minimal. Similarly, Barrett *et al.* (1996) comment that the findings of the Latham report in respect of project briefing were hardly different from those of Banwell 30 years previously.

That the Latham report did not address the fundamental problems of the industry was well expressed by Cox and Townsend (1997) who behaved that although well intentioned, the report was flawed, mainly due to the methodology adopted which relied on consulting vested interests within the industry that were intent on maintaining the status quo. They believed that by asking industry groups about problems they face each group identified their own problems and that an amalgamation of such views was unlikely to solve the problems of the industry, which are about constructing teams.

Subsequently, the Egan Report, *Rethinking Construction* (Egan 1998), focused on the lack of innovation and the lack of investment in research and training in the industry. The report argued for a radically changed industry with innovation from procurement to building systems, and with design and construction organised along manufacturing lines. The objectives were higher margins for contractors, better value for money for clients, improved welfare (particularly safety) and better training. Many of these benefits were seen to be achievable through supply chain management using long-term partnerships. The report identified five key drivers for change: 'committed leadership, focus on the customer, integrated processes and teams, a quality-driven agenda and commitment to people'. In every edition since its very first in 1984, this book has sought to show the value of systems theory in thinking about how to improve these elements, and this edition continues to do so. Such a view has more recently been taken up by others (Blockley & Godfrey 2000; Fewings 2005).

The effort which has gone into the follow-up to the Egan Report far outstrips anything which has gone before. The Strategic Forum for Construction (SFfC, 2006) was established to take its recommendations forward through the co-ordination of all the bodies associated with the industry. Its vision is the consistent delivery of world-class construction products. Whilst the Egan Report was much wider than previous reports and did not explicitly focus on project management, it is interesting that its four key areas are client engagement, integrating teams and supply chains, people issues and enhancing the value of the product – which comprise the essence of project management. The SFfC's (2002) major publication was *Accelerating Change*, which identified progress since the Egan Report, including innovation, key performance indicators and, most importantly, demonstration projects which 'provide the opportunity for leading edge organisations from whatever part of construction to bring forward projects that demonstrate innovation and change that can be measured and evaluated'. It noted that over a thousand construction organisations were actively involved in its initiatives – which in itself provides a major integration problem! SFfC has also set up a number of significant groups to further progress its agenda, including Constructing Excellence, the Construction Clients' Group and the Sustainability Forum. The long-term challenge is for the initiatives to percolate to all levels of the industry rather than remain with the more progressive, and usually large, firms.

Such reviews, reports and developments have been invaluable in identifying initiatives for modernising the construction industry but they have, by definition, focused on immediate and practical issues. Underlying many of the challenges and ideas which were identified was the need for a conceptual framework for designing organisational structures. Such a framework would allow project management functions to be identified so as to reflect the demands of different projects and enable initiatives appropriate to the specific project to be incorporated.

Relevance of systems theory

There have at last been some valuable initiatives in response to the pleas of the successive official reports for greater co-ordination, but there remains a resistance to change which reflects strongly entrenched attitudes and loyalties. Against such a background, progress has been incremental. However, an equally significant inhibition to progress has been the lack of a fundamental framework of organisation theory relating to construction projects against which experience of the various organisational initiatives can be measured and compared.

Systems theory provided the opportunity to develop such a framework. *General systems theory* (GST) originated in the biological sciences, but its originator (von Bertalanffy 1969) has acknowledged its general applicability, which he considers encompasses business organisations. It has been usefully applied to organisational problems in industries other than the construction industry.

The attraction of systems theory as a medium for identifying a conceptual framework for the management of the construction process lies in the basic premise that a system is an organised or complex whole: an assemblage or combination of things or parts forming a complex or unitary whole, which is greater than the simple sum of the parts. The systems approach stresses the contribution of the interrelationships of the parts of the system and the system's adaption to its environment in achieving its objective. A major outcome of the application of systems theory has been the acceptance that there is no universal panacea for management problems. Recognition of this by management theorists and researchers has led Jackson (2000) to dismiss management 'gurus' who claim to have unequivocal solutions to all management issues in all circumstances. As he says, 'Those who propagate management fads often appear to know the answers before they actually do the research, which amounts to little more than a self-fulfilling prophecy.' He quotes Lorsch's (1979) criticism of human relations theory in such terms and points out that Lorsch's strictures have been echoed more recently, almost to the letter, by Harvey-Jones (1993):

> 'The difficulty is that there can never be any single correct solution to any management problem, or any all-embracing system which will carry one through a particular situation or period of time . . . choosing the particular ideas which are most appropriate for the position and time in which he finds himself.'

Early recognition of the value of the application of systems concepts to the organisation design of the construction process was demonstrated by Morris, Napier and Handler. Peter Morris (1972) developed an approach to studying integration of the participants at the design–construction interface of construction projects. Morris's work supported the systems approach in that he found that organisation theory, especially when employed in the context of a systems framework, could be used to describe and explain the nature of the management process of construction projects.

A further application of the systems concept was made in Sweden by Napier (1970). In this work he attempted to gain an understanding of the problems of the Swedish building industry as a whole as a basis for the design of systems for the future. Handler (1970) was principally concerned with the building as a system. This concept was developed by reference to GST, by drawing an analogy between a living organism and a building.

Later work of significance was undertaken by, for example, Ireland (1985), Rowlinson (1988) and Hughes (1989). All were underpinned by a systems perspective applied to different issues of importance to the construction industry. Ireland used the systems model of organisations as a general paradigm and from that base investigated the relationship between the use of particular managerial actions and the achievement of project objectives. Rowlinson attempted to indicate some rational basis for the choice of procurement form for the management of construction projects by identifying those factors which significantly affect project performance. This was done with particular reference to the distinction between design-and-build and traditional procurement forms. Hughes adapted the model described in this book to analyse the management of public sector projects to identify the elements of the project management process which contributed or detracted from their success. Subsequently the systems approach has become accepted as a powerful tool for analysing construction project management (cf. Blockley & Godfrey 2000; Fewings 2005).

These studies illustrate the value of the application of systems theory to the building process. Each study took a different perspective, but employed the same basic concepts. The fundamental premise of systems theory stresses interrelationships and is as concerned with the links between the parts of the system as with the parts themselves. The problem of how to make the links work effectively is essentially the problem of project management. In order to apply these ideas to the construction process to the greatest benefit, it is necessary to take as broad a perspective of the process as possible from conception of the project to completion and even beyond.

Chapter 3
Organisation and the Construction Process

Introduction

In a business setting managers depend on effective organisations for achieving their objectives. But there are also organisations in many contexts other than business, for example public, social and sports organisations. In all cases the members seek their objectives through their organisations. Organisations are ubiquitous. It is hardly surprising therefore that studies of organisations have arisen from many traditions, from both practice and academia, and from such disciplines as sociology, psychology, economics and management and, more recently, the feminist movement.

For project managers the problem is to distil, from what was described by Koontz in 1961 as the 'management theory jungle' (and which remains a jungle today), those aspects of the work on organisation that are relevant and useful to managing the construction project process more effectively, particularly as the jungle has become much more dense since Koontz's description over 40 years ago.

The objective of this chapter is to briefly explore the range of perspectives on organisation and identify those attributes which are valuable in a project management context, before developing them further in future chapters. Classifying organisational approaches is fraught with difficulty as each phase does not have a distinct beginning and end. Rather, organisational approaches evolve over time as each new approach draws something from previous approaches, so the distinctions apparent in the presentation here should not be seen as absolute divisions.

McKinley *et al.* (1999) have studied the evolution of schools in organisation theory and suggest that:

'...a dynamic tension must exist between novelty and continuity in evolving organisation theory schools – and the school must also be of sufficient scope – for school based empirical research to be established. In organisation theory schools can fade away, seemingly independent of empirical validity, especially if they appear to hold little of novel value for scholars. And profound revolutions are not typical; rather schools gradually become more articulated.'

The classical approach

The classical approach stems from the first wave of writing on management in the early part of last century and is characterised by the work of Taylor (1911) and Fayol (1949 trans.) and was the foundation of management practice up to about 1950, although many will say that it still continues to be so in many cases today.

Taylor laid the foundation of 'scientific management'. The approach determined that it was possible to scientifically analyse and structure the tasks to be performed so that the maximum output could be obtained with the minimum input. This approach meant that people were perceived as machines and efficiency was the sole criteria of success. The outcome of such an approach led to increasing specialisation of the workforce. Managers' activities were also seen to be governed by set processes and procedure as much as the workers.

Fayol, whose work did not become freely available until after translation in 1949, was nevertheless influential before then as others in the USA had developed his ideas. They developed 'principles of management' which were concerned with such things as pyramidal structure, unity of command, line and staff, the scalar chain and span of control. The primary element was the pyramidal organisation structure and the idea that authority is delegated downwards. Division of labour was advocated so that the sub-goals of the various units would add up to overall organisational goals and co-ordination would be handled through the management hierarchy. The principles emphasised formalisation and specialisation and were in this way complementary to and supportive of Taylor's scientific approach. It may be that in those days the construction professions recognised, through their experience, that such approaches were not really appropriate to the management problems of construction projects and, for this reason, did not fully develop management in their training at an early date.

What emerged from this classical view of management and hence organisations was a deterministic perception. The 'principles' were held to be universal truths about how management should be undertaken and the only way to manage business activities or processes. Hence an extremely rigid view of how to organise emerged. As with many originators of theory, Fayol did not intend this outcome. It was the users who followed his 'principles' slavishly (and many still do so today). Fayol made the point that they were not rigid. Maybe his mistake was calling them 'principles':

'The soundness and good working order of the body corporate depends on a certain number of conditions termed indiscriminately principles, laws, rules. For preference I shall adopt the term principles whilst disassociating it from any suggestion of rigidity, for there is nothing rigid or absolute in management affairs, it is all a question of proportion. Seldom, do we have to apply the same principle twice in identical conditions; allowance must be made for different changing circumstances.' (Fayol 1949 trans.)

His contemporaries made the point even more tellingly:

> 'Students of administration have long sought a single principle of effective departmentalization just as alchemists sought the philosopher's stone. But they have sought in vain. There is apparently no one most effective system of departmentalism.' (Gulick & Urwick 1937)

It has taken many years for these early signals to be taken on board by both practitioners and academics but developments in organisational theory have eventually reflected these views.

Of great influence at this early stage in the development of thinking on organisations was the work of the German sociologist Max Weber (1968 trans.) He worked independently of Taylor and Fayol and adopted a different stance. Rather than focusing on how to improve organisations he took a far more academic approach of seeking to describe the characteristics of newly emerging bureaucratic structures. The characteristics he identified were generally compatible with other ideas of the time such as specialisation, hierarchy, etc., but were developed in much more depth. His primary focus was on organisations as power structures in which control is achieved through an organisation hierarchy. Discipline was the keyword which required the exact execution of orders from above.

The major writers within this approach represented different schools. Taylor was highly pragmatic, basing his work on scientific planning. Fayol was from the administrative theory group, which was prescriptive in that they believed in principles of management applicable to all situations, whilst Weber was concerned with describing bureaucratic structures. All this led to Bennis (1959) describing the classical approach as representing 'organisations without people'.

The classical approach to organisations and management was therefore seen as essentially rigid and originated from military and church models which strongly influenced the way in which the early managers organised. It did not make explicit the effects of the human component and external influences on organisations. Instead, focus was entirely on the internal characteristics of the organisation.

This is not to say that all the ideas of traditional management have been (or can be) discarded by modern methods of management. For example, the need for hierarchy in ordering and controlling organisations is very much part of the classical approach. However, the general reaction to hierarchies is that they are oppressive and not what people want; they are also seen as out of date, inefficient and inhumane. But hierarchies continue to play a major role in modern organisations. The reasons for their continued existence are well argued by Leavitt (2005) as ambition, power, status, and wealth as well as devices for coping with complexity. Hence students of modern organisation and management cannot afford to ignore earlier management thinking.

Although seen to be an historical aspect of management by theorists, Green (1998) highlighted that the classical approach has re-appeared under the guise of business process re-engineering (BPR). He believes that within the construction literature there has been uncritical acceptance of BPR as 'a radical new beginning' as claimed by Hammer and Champy (1993). He points out that few, if any, central concepts of BPR are new and that several authors have suggested that BPR is a return to the simplistic machine metaphors of Taylor's (1911) concept of 'scientific management'.

The behavioural approach

Serious study of people in organisations did not begin until it was explicitly recognised that informal organisations existed in parallel with formal organisations. Recognition of informal organisation structures alongside the formal, and the shortcomings of classical organisational theory, saw the emergence of the behavioural schools, which believed that the study of management should be centred on interpersonal relations or that it should be seen as a social network.

Informal structures exist alongside formal organisational structures because people cannot be treated as machines. Their behavioural responses to their position within a formal organisation cannot be expected to subscribe to the predetermined manner in which they are expected to perform. Hence an informal structure will arise. How different this structure is from the 'official' structure will depend on many factors not least of which will be how well the formal organisation has been constructed. Contributing significantly to this scenario is goal complexity within organisations. There is often a disparity between the official goals of an organisation and the goals actually pursued which then govern the behaviour of the participants. In such situations the existence of unofficial goals creates an informal structure intended to achieve them. Hence the behaviour of the members of an organisation cannot be constrained within an inappropriate organisation structure. The recognition of this phenomenon challenged the classical approach to organisations. A wide range of researchers contributed to the development of this new approach, far too many to cover here, but a few important ones are briefly discussed to give a sense of their thinking.

The famous Hawthorn experiments conducted by the Harvard Business School in the 1920s and 1930s laid the groundwork for much that followed. As Scott (1992) states:

> 'The experiments served to call into question the simple motivational assumptions on which the prevailing rational models rested. Individual workers do not behave as "rational" economic actors but as complex beings with multiple motives and values; they are driven as much by feelings and sentiments as by facts and interests; and they do not behave as individual, isolated actors but as members of social groups exhibiting commitments and loyalties stronger than their individualistic self-interests.'

As a result, many sociologists and social psychologists devoted their attention to examining how people in organisations actually behaved, how they actually related to their supervisor, subordinates and peers and what were the factors which motivated members of organisations. McGregor's (1960) work encapsulated much of this work and contrasted the social system approach with the classical approach through his now famous 'Theory X' and 'Theory Y' assumptions about how people behave in organisations. Assumptions underlying Theory X were:

- 'Individuals dislike work and will seek to avoid it.'
- Therefore, 'most people must be coerced, controlled, directed, threatened with punishment to get them to put forth adequate effort toward the achievement of individual objectives'.
- 'The average human being prefers to be directed, wishes to avoid responsibility, has relatively little ambition, wants security above all.'

By contrast, under Theory Y:

- Most individuals do not 'inherently dislike work . . . the expenditure of physical and mental effort in work is as natural as play or rest'.
- 'External control and threat of punishment are not the only means for bringing about effort toward organisational objectives.'
- The most significant rewards are those associated with 'the satisfaction of ego and self-actualization needs'.

Barnard's (1938) significant earlier work also falls within this general category. He stressed that organisations are co-operative ventures which integrate the contributions of their members. He also dealt significantly with authority in organisations (referred to in Chapter 8) which is reflected in his view that goals are imposed from the top down whilst their attainment depends on the bottom up. Much of the work on leadership (see Chapter 9) also stemmed from the behavioural approach.

The behavioural movement began in the 1930s and continued into the 1970s with perhaps its zenith in the 1950s. However, questions were raised as to whether the field of human behaviour is equivalent to the field of management as the nature of an organisation's goals and consequential necessary tasks of organisations can make this approach impracticable. Also research on the basis of empirical evidence has failed to show relationships between behavioural aspects of organisational members and productivity (Scott 1992). Eilon (1979) refers to the 'myth of self-actualisation' and Bennis (1959) describes this approach as being about 'people without organisations'. In common with the classical approach's normative view of organisations, the behavioural approach does not take into account the environment in any consistent manner but generally treats organisations as closed systems.

Between and within the classical and behavioural schools are a wide range of approaches. The main criticism now levelled at them is that at the time each was offered as the one best way to organise. Subsequent organisational structure thinking denies such an assumption but believes that each may have something to offer within a systems framework.

The socio-technical approach

The Tavistock Institute of Human Relations, London, undertook a series of studies in the 1950s and 1960s which contributed significantly to the development of systems theory and its application to business organisations and the construction industry.

They developed what at the time was a distinctive research approach in that they proposed that the unique feature of business organisations is that they are both social and technical systems. The socio-technical approach emphasises that the needs of both the technical and social aspects should be served by organisations. This view contributed to combining many of the previous approaches some of which considered only technical needs whilst others considered only social needs. As previously mentioned, Bennis (1959) labelled the former 'organisations without people' and the latter 'people without organisation'. Scott (1992) notes that the goal should be one of 'joint optimisation' of the needs of both the technical and social systems, since the two systems follow 'laws' and their relationship represents a 'coupling of dissimilars' (Emery 1959). From this view emerged the recognition of the impact of environments on organisations. Scott (1992) also notes that the Tavistock work is essentially European with little similar work having been carried out in the USA.

The Tavistock group undertook an important study of communications in the construction industry (Higgins & Jessop 1965; Tavistock Institute 1966) which identified the main features of the technical system as 'interdependency and uncertainty'. In terms of the social system they highlight the mismatch between the traditional organisational arrangement with the architect as designer/manager and the organisational separation of production undertaken by the construction company. Their report was the first to identify the need for someone in a separate project management role. Winch (1989) criticises their work on the grounds that 'nowhere in either of the two reports is there any analysis of the implications of the contracting relationship . . .' and identifies two weaknesses: 'It was developed for handling relationships within simple organisations' and '. . . the perspective is psychologically orientated, and so has difficulty in grasping a context in which the actors have differing economic interests'. Winch puts these arguments within the context of supporting a transaction cost approach to the construction firm and the construction project.

What Winch fails to do in these arguments is to separate the firm from the project. The Tavistock group was concerned with process and the needs of the fundamental organisational system for the production of a project which was satisfactory to the client. The manner by which the organisational units which carry out that process are formed is then a separate matter and the transaction cost approach may well be an elegant explanation of this structure. Nevertheless the process of creating and constructing a project has basic organisational needs which should be satisfied by the provision of an ideal organisational arrangement which a client should seek in order for their project to have the best chance of satisfying their needs.

This argument is put here as it is a recurring issue in the application of organisational theory to construction projects. The needs of the management of the process and the needs of the management of the firms should in the first instance be considered separately but with the needs of the process taking precedence so that the selection of organisational units can be taken in an informed manner. That is not to say that Tavistock should not have taken transactions costs into account, although much of the work on the link of the transaction cost approach to organisations took place much later than Tavistock's pioneering work.

The systems approach

Interestingly, the reasons for the emergence of these two different views of organisations (i.e. classical and behavioural) have been attributable to variations in the background of the authors (Lawrence & Lorsch 1967) and to the fact that each approach concentrated on different types of organisation (Scott 1998). In the former, practical men v. academics, and in the latter, industrial firms and state bureaucracies v. service and professional organisations. Lawrence and Lorsch also add that the two types of organisation were operating in different types of environments. Scott (1998) believes that the two approaches are characterised by divergent views of the fundamental nature of social systems. All of which has meant that the classical and behavioural approaches did not seem to offer much to the construction professions and industry, although construction firms may have been attracted to the formality of the classical approach. As a result, traditionally there was little management included in courses for the construction professions but rather more in courses directed at the construction firms. In particular the classical and social systems schools did not appear to have anything to offer the management of the process of producing a project. The managers of the process were alienated by each school claiming to offer the only way to manage.

The management abilities of the professions and industry were, therefore, mainly acquired through the experience of managing in the real world. Although such experience is vital, its value is reduced if it cannot be gained within a conceptual framework of management theory. The ad hoc acquisition of theory has meant that generally most older professionals' and industrialists' knowledge relates to traditional management concepts. It was only with the advent of the systems approach to organisation structure that members of the construction professions and industry had a theory to which they could relate their experience of their particular industry.

The systems approach is essentially a way of thinking about complex processes so that the interrelationships of the parts and their influence upon the effectiveness of the total process can be better understood, analysed and improved. Its origins lie in the biological sciences through its founder Ludwig von Bertalanffy who devised general systems theory (GST) from his consideration of the fundamental interdependency of many aspects of science which were studied independently. He generalised his theory (von Bertalanffy 1969) to show that it was applicable and valuable to a broad spectrum of disciplines and it was subsequently applied to business organisations.

The appeal of the systems approach to the study of construction project organisations arises from its focus on how the parts of a process are dependent upon each other, as illustrated by the following definition of a system:

> 'An entity, conceptual or physical, which consists of interdependent parts. Each of a system's elements is connected to every other element, directly or indirectly, and no sub-set of elements is unrelated to any other sub-set. (Ackoff 1969)

It is clearly the case that the success of the construction process depends to a large extent upon the way in which the architect, engineer, quantity surveyor, contractors and others work together. It depends upon them perceiving the same objectives for the project and recognising that what each of them achieves depends upon what the others do. With this view they should be able to stand above the particular interests of their own contribution and see the problem posed by the project as a whole. The advent of the project manager has, to a large degree, come about as a result of the inability of the contributors to consistently achieve this, and in response to the consequent need for someone to concentrate solely upon integrating the various contributors in the interests of the client.

To understand how the building process operates as a system it is necessary to understand the distinction between closed and open systems. *A closed system* is one that does not respond to events and occurrences outside the system. It cannot adapt to changes and is therefore predictable. Machines can be considered to be closed systems in that the parts are selected to perform specific functions in a given set of conditions to produce a predetermined output. If there are changes in the conditions for which the machine was designed, the machine will not adapt to them. For example, a washing machine will not work if overloaded, a motor car will not work properly on dirty petrol.

On the other hand, an *open system* adapts to events and occurrences outside the system. These events and occurrences take place in what is known as the system's environment. This has been defined as a set of elements and their relative properties, which elements are not a part of the system but a change in any of which can produce a change in the state of the system (Ackoff 1971). An open system has a permeable boundary and there is import and export between an open system and its environment. It is therefore influencing and being influenced by its environment. An open system is dynamic and adapts to its environment by changing its structure and processes. Although stable, it is always changing and evolving and presents differences over time and in changing circumstances. A living organism is an open system and business organisations are analysed as open systems.

However, it is not as clear cut as the closed–open dichotomy implies: there is a range of other classifications. For example, a central heating system and the human body, both of which adapt themselves to changes in the temperature of their environment by internal adjustment so that they remain static, are referred to as homeostatic systems. Also, Child (1977) has described a system that exists

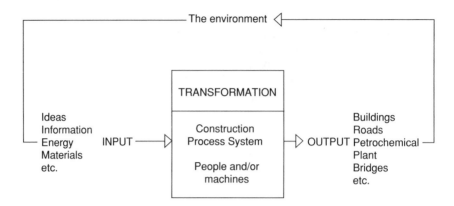

Fig. 3.1   The construction process as an input–output model.

in a protected environment in which it defends itself from having to adapt fully to its environment. Therefore the system is not fully open.

Business organisations could never have existed as closed systems. Similarly, the construction process has always been an open system. Potential clients exist in the environment of the construction process system and the system must adapt to them. It imports ideas, energy, materials, information, etc., from its environment and transforms them into its output, which is the finished construction. This is then exported to the environment, which is itself influenced by the use to which the completed project is put and by the fact that the construction is an addition to the nation's fixed capital. The process is illustrated in Fig. 3.1.

Recognising the construction process as an open system means that the functions upon which the project management process should focus can be summarised as:

- identifying, communicating and adapting the system's objectives
- ensuring that the parts of the system are working effectively
- ensuring that appropriate connections are established between the parts
- activating the system so that the connections that have been established work effectively
- relating the total system to its environment and adapting the system as required in response to changes in its environment.

In practical terms the project manager will be concerned particularly with anticipating the chain reactions of decisions and developments that occur on the project. For example, as a result of an upturn in business, the client may decide at a late stage in the design of a project to be submitted for competitive tender that substantially more floor area is needed in the factory. This decision has to be appraised in terms of its effect on the project cost, completion time and functional efficiency, and evaluated against alternatives such as providing the additional area in a different form, e.g. leased accommodation, or by a different method, e.g. a negotiated contract for the additional area. This will require interaction between all the contributors to the project, the complexity of which will depend upon the actual stage of development of the project. The final decision will need to be taken solely in terms of the client's objectives in relation to the revised requirements. For example, completion time not only means additional construction time but also the additional time required by the consultants and the effect this may have on construction completion, particularly if this may mean 'hidden delays' to construction completion because of drawings being issued late or incomplete. The relationship between cost and completion time would also need to be appraised. All of these factors would require interaction between the contributors so that the priority of the competing demands would be resolved in the client's interest. The project manager needs to be able to anticipate the interconnectedness generated by such decisions and to manage the system with respect to them.

General systems theory developed alongside the various schools of management thought and it had an attraction for management thinking as it presented an opportunity to converge these strands of thought within an acceptable and theoretically sound framework with less rigidity and more recognition of interdependency in organisations than previous approaches allowed. The systems approach reflects the scale of interdependency created by the nature of activities to be undertaken (e.g. the design and construction of a building) and the effects upon the activities of environmental influences. It therefore discounts rigid approaches that propose one method for all circumstances. This is not to say that the systems approach discounts as irrelevant the ideas of classical management and the behavioural schools (as illustrated earlier in the brief discussion of hierarchies), which are still pervasive in practically all organisations today whether designed on the basis of the systems approach or not; but rather that it provides a framework for understanding and analysing organisations through their internal and external relationships, which places into context the earlier views of organisations. For example, the behaviour of individuals within an organisation remains important but it is more easily understood and relevant if it is seen within the context of the relationships demanded by the activities being undertaken and the environment within which they take place.

Such perspectives were recognised in the early application of systems theory to organisations. As Scott (1998) points out, Etzioni (1964) insisted that all organisations exhibit conflicts between formal and informal structures, rational and non-rational aspects of behaviour, and controlling and controlled participants. Also Thompson (1967) is correct in saying that some parts of an organisation are more protected from environmental influences and some parts more governed by classical (rational) influences whilst others are more influenced by behavioural (natural) phenomena. He goes on to argue that the straight time line from classical to behavioural to systems approaches is a simplification and proposes a layered model of dominant theories which leads to four periods; closed-rational (1900–1930); closed-natural (1930–1960); open-rational (1960–1975); and open-natural (1975–present). He places contingency theory and transaction cost analysis into the open-rational category and socio-technical systems, resource dependency and institutional theory into the open-natural category, all of which are described later and from which it can be seen that construction projects sit most comfortably in the open-rational category.

Systems thinking can be applied more broadly than just to organisations. Its scope is illustrated by Jackson (2000) by reference to Lane and Jackson (1995), who provided an annotated bibliography reflecting the 'breadth and diversity of system thinking'. They identified eight strands as general systems thinking, organisations-as-systems, hard systems thinking, cybernetics, system dynamics, soft system thinking, emancipatory systems thinking, and critical systems thinking. He then goes on to apply systems thinking to four types of social theory identified as the functional, interpretive, emancipatory and post-modern systems approaches. Jackson's work further illustrates the ubiquitous nature of the application of systems thinking by referring to 'the various professional societies and academic groupings which advocate system thinking, and the periodicals, newsletters etc., that promote systems thinking, as well as those who research and practice using a systems approach'. He continues, 'Systems thinking is then a general term used to denote the theories, methodologies, models, tools and techniques, which are based on systems ideas and concepts and are employed by those who argue for a systems approach'.

It can be seen therefore that the approach to construction project management taken here is part of a family of ideas arising from the systems approach. Not only is the systems approach ubiquitous in its application, it is also ubiquitous within its application to organisations by its ability to accommodate other management approaches.

Contingency theory

Impetus to the application of the systems approach to management came through Lawrence and Lorsch's (1967) major study which led to the contingency theory of organisations design, which states that there is no one best way to organise but rather that organisation is a function of the nature of the task to be carried out and its environment. It encompassed many applications of systems ideas to organisations. Lawrence and Lorsch found that different environments, which generate different levels of uncertainty, require varying degrees of separation (differentiation) of organisational units (e.g. architect, engineer, contractor) and hence they require different degrees of integration.

The extent of differentiation within an organisation depends upon the uncertainty and diversity of the environment and the effect this has on the way the task has to be organised and managed. Lawrence and Lorsch state that they found that the amount of differentiation in the effective organisation was consistent with the environmental demand for the interdependence of the parts of the organisations. In developing their contingency theory they state that this starting model is complicated as soon as we move to a complex, multi-unit organisation, in which each unit strives to cope with different parts of the environment. For example, a construction project that is carried out in conditions of uncertainty and is technologically complex requires a wide range of specialist skills, which are closely dependent upon each other, in achieving a successful outcome. As soon as this happens, it introduces the complication of integrating the work of different units. Lawrence and Lorsch see the existence of an integrating unit and conflict-resolution practices as contributing to the quality of integration and in turn to overall performance. This unit has come to be represented on construction projects by project managers.

A number of other significant research studies building on systems theory led up to the contingency theory, for example, that by Burns and Stalker (1966) which analysed firms in the electronics industry and identified two patterns of organisations and management. The one they termed 'mechanistic' was similar to the classical model referred to earlier. The other, termed 'organic', had a participative character. The 'mechanistic' and 'organic' structures lie at the extremes of a spectrum which illustrates the range of approaches possible. Burns and Stalker did not suggest that either was superior to the other. They concluded that, when taken in context with the task and environment being considered, one pattern will be more appropriate for the specific tasks and environment in question.

The contingency theory is a succinct summary of a great deal of detailed work that went before it. It is perhaps a reflection of the management discipline's apparent need to sum up a complex situation in just a few words. Child (1977) is critical of the contingency theory on these grounds and believes that it has not in the main recognised the organisation design difficulties which may result from the presence of multiple contingencies. He is concerned at the situation in which a configuration of different contingencies is found which are conflicting in terms of organisation design. For instance, a construction project may demand a relatively bureaucratic organisation structure to ensure accountability, but at the same time require a more loosely structured organisation to more readily allow innovation to take place. Child also questions the cost effectiveness of the additional integrating mechanisms required, as he is not convinced that there is evidence that they improve performance.

Scott (1998) in his case for contingency theory being an open-rational (classical) type asks how, given that an organisation is open to the uncertainties of its environment, can it function in a classical (rational) manner? He argues that Thompson (1967) believes that it can do so by creating some closed system compartments in critical parts of its structure and details how this can be achieved.

Nevertheless, even allowing for such criticisms, the systems approach as summarised in the contingency theory provides a framework for thinking about the design of construction project organisations and for analysing them, so that the effect of organisation structure on the outcome of projects can be better understood.

Strategic contingency

The strategic contingency approach adopts an open system approach as does contingency theory but it arrives at a different rationale for the structure of organisations. Contingency theory believes that managers have to respond to the environment of their organisations in designing organisations hence they are responsive to, and their actions determined by, the environment. Strategic contingency theorists believe that managers have choices (Child 1972) and although the environment may constrain their choices to some extent it does not determine them. They recognise the role of power in determining the strategy to be adopted.

Following from this view is that, rather than being a function of task and environment, organisation structures are determined by political contests within organisations (Pfeffer 1978) leading to a framework for the power-driven political explanation of organisational structure.

Resource dependency

The resource dependency model also arises from the open system framework and can be seen to be associated with the strategic contingency approach as its primary concern is the impact of external forces on how firms organise (Pfeffer & Salanick 1978). There are two major elements. One is that organisations are constrained by and depend on other organisations that control resources which are critical to their operations and the other is that organisations attempt to manage their dependencies on external groups to acquire more autonomy and freedom. The resource dependency model sees managers making strategic choices within constraints to reduce their dependencies, which illustrates the model's similarity to the strategic contingency approach. However, the model's view is that managers do not have unbridled strategic choice as Child (1972) originally proposed in developing the strategic contingency approach but that they do exercise some discretion over how to structure organisational relationships to manage the uncertainties created by dependency which requires adjustment of inter- and intra-organisational linkages as summarised by Greening and Gray (1994).

Scott (1992) believes that the resource dependency model means that organisational participants, particularly managers, scan the relevant environment searching for opportunities and threats, attempting to strike favourable bargains and to avoid costly entanglements. All organisations are dependent on supplies and consumers but which specific exchange partners are selected and what are the terms of exchange are partly determined by the organisation itself. Astute managers acquire the necessary resources but do so without creating crippling dependencies.

Institution theory

Institution theory is concerned with the impact of institutions (government, professional groups), public opinion and pressure groups on the structure of organisations. Meyer and Rowan (1977) argue:

'Many of the positions, policies, programs, and procedures of modern organisations are enforced by public opinion, by the views of important constituents, by knowledge legitimated through the educational system, by social prestige, by the laws, and by the definitions of negligence and prudence used by the courts. Such elements of formal structure are manifestations of powerful in-stitutional rules which function as highly rationalized myths that are binding on particular organisations.'

Institution theory is part of the group of ideas which stem from the treatment of organisations as open systems. It adopts a perspective which is different from the strategic contingency and resource dependency approaches in that its focus is that many of the more powerful forces in the environment are social and cultural pressures to conform to institution preferences and conventional beliefs rather than pressures to adapt to a more productive organisation. Response to such pressures is seen to provide organisations with increased legitimacy, resources and survival capabilities. Nevertheless, other researchers suggest that organisations should also attend to the demands of their technical environments (Meyer & Scott 1983).

Population ecology

Along with institutional theory, population ecology is a most prolific school of North American organisational theory. Whilst the main focus of the theories of population ecology is not organisation structure per se, its focus on explanations of the reason for variations in organisations and the patterns of their birth and death has much to say about organisation structure. However, it does not see structure as mandated by efficiency criteria. It operates therefore at a level of focus and abstraction somewhat removed from those organisational theories of more immediate concern to project management; but 'at its heart it has a concern with structure, with the assumption that, in understanding organisational births and deaths and the evolution of organisational forms, the concept of structure is critical' (Hinings 2003). A comprehensive overview of population ecology is presented by Aldrich (2000).

Criticisms of the systems approach

Criticisms of the systems approach to organisation stem from its supposed dis-regard of the informal organisation. It is argued that the systems approach as-sumes that people in the organisation 'act in a calculable and predictable man-ner'. It is further argued that as systems thinking comes from natural science models, which do not have 'free-willing subjects and the consequent uncertainty and open-endedness of social behaviour', it can lead to mechanistic and overly rationalist ways of viewing organisations (Fincham & Rhodes 2005). But these criticisms hold in practice only if those in the position of managing the organisa-tion choose to believe that people in organisations behave mechanistically. Those involved in managing any aspect of construction know that that is not the case in their industry, which appears to attract the more extrovert.

The systems approach emphasises adaptability and this occurs in response to the system's environment which, in the case of construction, is transmitted to the system via its people who react to information and interpret it according to their characteristics. Viewing the organisation structure as system not only provides a sound representation of the process of project management but also assists in visualising the manner in which the people are performing within the system. This may be easier to achieve in construction project management structures than with other applications, due to the specific nature of the professional tasks involved in the process.

A further criticism of systems application to organisations is the idea of an organisation having a goal or single objective. This is seen to be oversimplified. Firstly, organisations do not have goals; people in organisations have goals. Secondly, there are likely to be multiple goals in most organisations. These issues are likely to be less severe in construction as the main task of constructing the project is clear. Whether the goal is so clear in the client organisation is another matter; and if it is not, this can lead to difficulties for the project team in achieving the client's goal. However, there may be many individual sub-goals both in the client organisation and within the project team.

Whilst criticisms may exist, the systems approach does provide a framework for understanding organisations that can incorporate other aspects of organisation theory. Scott (2003), whilst referring to rational and natural systems, makes the point that they are complementary and that each view represents a partial truth: 'If the perspectives seem at times to conflict, this is because the organisational elements to which they point sometimes conflict. The recognition of the inevitability of such conflict is an important part of the "whole" truth about organisations'. This is a sentiment which can be seen to apply across the full range of organisation theory.

Criticisms of contingency theory centre on the claim that it has oversimplified the problem of organisation design. That is, the reality of management is far more complex than contingency theory acknowledges, and fitting the organisation structure to the prevailing contingencies to produce better results is not a practical proposition. In addition, as discussed earlier, strategic contingency theorists criticised contingency theory as incomplete, as they proposed that managers have power, which gives them choices.

Donaldson (1996) has been a prime defender of contingency theory. He argues for positivist organisation theory, stating:

'Structural contingency theory is positivist and functionalist. Critics have expressed doubts about both of these attributes and have striven to develop analyses of organisations that are anti-positivist and anti-functionalist. These theories include strategic choice theory, political theory, typology theory and organisational systematics. However none of these anti-positivist or anti-functionalist theories that have been considered here are sustainable. They suffer damaging problems at either the theoretical or empirical levels – and frequently at both levels.'

Fincham and Rhodes (2005) agree with Donaldson, saying: '...the contingency theorists' notion of decision making as an activity partly constrained, but displaying choice and variability, does not seem unreasonable'. Whilst contingency theory may have lead to unrealistic expectations that research would clearly show that structural design made to tightly fit environmental conditions would produce higher performance (Jackson 2000), nevertheless it has proved to be an eminently practical approach to thinking about the analysis and design of organisation structures.

The enduring nature of systems theory is reflected in its continuing application to organisations more than fifty years after its emergence as a significant way of perceiving organisations (Ruegg-Sturm 2005). Its application to construction project management organisations in this book has been particularly productive, reflecting the necessity for differentiation and the consequent importance of integration required to cope with increasingly challenging environments (Winch 2002). Nevertheless, Scott's (2003) recognition that the existence of multiple paradigms may 'reflect the complexity of the phenomenon addressed and improve our analytical capacity by providing multiple lenses though which to observe the world' is a valuable observation.

Mintzberg's classification

Typologies and taxonomies are of continuing importance to the study of organisation structures and had a revival in the 1980s and 1990s (Hinings 2003). A foremost example of a typology in organisation theory is that of Mintzberg (1979, 1989).

Mintzberg's classification of the structure of organisations has a strong appeal to people in the construction industry as demonstrated by Bennett's (1991) adaptation of Mintzberg's (1989) idealised types to suit construction, to which further reference will be made later.

Mintzberg's work is based on an open systems approach incorporating contingency theory as he believes that effective organisations achieve an appropriate balance between task, environment and organisation structure but he sees his configuration approach taking it further. This he characterises as 'getting it all together', in which the elements are selected to achieve consistency. He argues that academic research on organisations has favoured analysis over synthesis and has focused on how variables arrange themselves along linear scales rather than how attributes configure into types.

His basic premise is that a limited number of configurations can help to explain much of what can be observed in organisations. His analysis of organisations prior to synthesis is extensive and insightful but there is no space to do it justice here and the reader is referred to the original work. His seven configurations do, he believes, encompass and integrate much of what is known about organisations. He emphasises that each configuration is idealised. As he says: 'a simplification, really a caricature of reality. No real organisation is ever exactly like any one of them. But some do come remarkably close, while others seem to reflect combinations of them, sometimes in transition from one to another'.

It should be pointed out that five of his configurations appeared in his book *The Structure of Organisations* (Mintzberg 1979) – Entrepreneurial, Machine, Diversified, Professional and Innovative. The other two – Missionary and Political – were added later (1989) and are distinctly different from the earlier five in that they overlay on the other five and are seen as forces in organisations rather than the form of an organisation. However, in their extreme forms they can become so strong that the organisation's structure is built around them.

The seven configurations follow using Mintzberg's own words. In the case of the original five he also includes structure, context, strategy and issues in their specification for which the original text should be consulted:

The Entrepreneurial Organisation
Nature:

- 'Simple organisations that are run by their leaders.'

Conditions:

- 'Fostered by an external context that is both simple and dynamic. It must be relatively simple (say retailing food as opposed to designing jet aircraft) in order for one person at the top to retain so much influence, and it is the dynamic context that requires the flexible structure, which in turn enables this organisation to outmaneuver the bureaucracies.'

The Machine Organisation
Nature:

- 'Supposedly the big bad guys of the organisation world, the homes of red tape and the sources of curious tales. Yet if we think of McDonald's or the Swiss railroad, a different impression develops, of organisations – when they get it right – that can be enormously efficient and can provide an unmatchable reliability of service.'

Conditions:

- 'Work of a machine bureaucratic nature is found, above all, in environments that are simple and stable. The work associated with complex environments cannot be rationalised into simple tasks, and that associated with dynamic environments cannot be predicted, made repetitive, and so standardised.'

In addition, the machine configuration is typically found in mature organisations, large enough to have the volume of operating work needed for repetition and standardisation, and old enough to have been able to settle on the standards they wish to use. These are the organisations that have seen it all before and have established standard procedures to deal with it. Likewise, machine organisations tend to be identified with technical systems that regulate the operating work, so that it can easily be programmed.

The Diversified Organisation
    Nature:

- 'The waves of mergers that have taken place over the last century have led to the formation of giant corporations and to the so-called divisionalised forms of structure. The 'conglomerate' is, of course, the ultimate example of this, where a corporation doesn't much care about any relationship among its different businesses other than financial.'

    Conditions:

- 'While the diversified configuration may arise from the federation of different organisations, which come together under a common headquarters umbrella, more often it appears to be the structural response to a machine organisation that has diversified its range of product or service offerings. In either case, it is the diversity of markets above all that drives an organisation to use this configuration.'

The Professional Organisation
    Nature:

- 'It is the one place in the world where you can act as if you were self-employed yet regularly receive a paycheck. These seemingly upside-down organisations, where the workers sometimes appear to manage the bosses, are fascinating in the way they work. We need professional organisations to carry out highly stable tasks in society, such as replacing someone's heart or auditing a company's books. But as a society we have yet to learn how to control their excesses: professionals who mistreat their clients, professional organisations that mistreat their supporters.'

    Conditions:

- 'The professional form of organisation appears wherever the operating work of an organisation is dominated by skilled workers who use procedures that are difficult to learn yet are well defined. This means a situation that is both complex and stable – complex enough to require procedures that can be learned only through extensive training yet stable enough so that their use can become standardised.'

The Innovative Organisation
    Nature:

- 'Some people refer to this type as 'high technology', and to its basic orientation as 'intrapreneurship', an indication that whereas the entrepreneurial configuration innovates from a central individual at the top, this one depends on a variety of people for its strategic initiatives. These initiatives tend to be many, because what adhocracy provides is sophisticated innovation. That comes at the price of a good deal of disruption, if not chaos, and wasted resources. This type achieves its effectiveness by being inefficient. Perhaps that explains why it confuses many people: the innovative organisation may be necessary, but it is not conventional, at least not by the standards of the traditional literature of management.'

Conditions:

- 'This configuration is found in environments that are both dynamic and complex. A dynamic environment, being unpredictable, calls for organic structure; a complex one calls for decentralised structure. This configuration is the only type that provides both. Thus we tend to find the innovative organisation wherever these conditions prevail, ranging from guerrilla warfare to space agencies. There appears to be no other way to fight a war in the jungle or to put the first man on the moon.'

The Missionary Organisation (Ideology)

- 'Rich system of values and beliefs that distinguishes an organisation.
- Rooted in sense of mission associated with charismatic leadership, developed through traditions and sagas and then reinforced through identifications.
- Can be overlaid on conventional configuration, most commonly entrepreneurial, followed by innovative, professional, and then machine.
- Sometimes so strong that evokes own configurations.
- Clear, focused, inspiring, distinctive mission.
- Co-ordination through the standardisation of norms ('pulling together'), reinforced by selection, socialisation, and indoctrination of members.
- Small units ('enclaves'), loosely organised and highly decentralised but with powerful normative controls.
- Reformer, converter, and cloister forms.
- Threats of isolation on one side, assimilation on the other.'

The Political Organisation

- 'Means of power technically illegitimate, often in self-interest, resulting in conflict that pulls individuals or units apart.
- Expresses itself in political games, some coexistent with, some antagonistic to, some that substitute for legitimate systems of power.
- Usually overlaid on conventional organisation, but sometimes strong enough to create own configuration.
- Conventional notions of concentrated co-ordination and influence absent, replaced by the play of informal power.
- Dimensions of conflict – moderate/intense, confined/pervasive, as well as enduring/brief – combine into four forms: confrontation, shaky alliance, politicised organisation, complete political arena.
- Can trace development of forms through life cycle of impetus, development, resolution of the conflict.
- Politics and political organisations serve a series of functional roles in organisations, especially to help bring about necessary change blocked by legitimate systems of influence.'

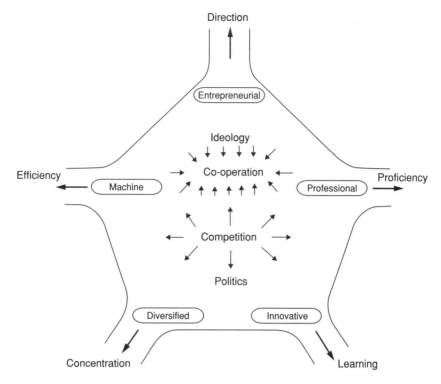

Fig. 3.2 An integrating pentagon of forces and forms. (*The Structuring of Organizations* by H. Mintzberg, © 1979. Adapted by permission of Prentice-Hall, Inc., Upper Saddle River, NJ)

Mintzberg then took his ideas on configurations a stage further by conceiving that, rather than being a perfect fit, many organisations fit *more or less* into one of his configurations. Organisations which would be expected to be a perfect fit for one of the organisational types demonstrated anomalies, some quite limited but others quite extensive. As a result he conceived his configurations as forces which exist to a lesser or greater degree in all organisations at one time or another if not all the time. At the same time he maintained that the seven configurations demonstrated the most fundamental forms that organisations can take and which some do for some of the time. This is summarised in Mintzberg's Pentagon of Forces and Forms shown in Fig. 3.2.

Mintzberg believes that the forces and forms as conceived by the pentagon 'appear to constitute a powerful diagnostic framework by which to understand what goes on in organisations and to prescribe effective changes in them'.

Bennett (1991) creatively uses Mintzberg's ideas and applies them to construction to provide valuable insights. He identifies three systems of organisation which he describes as idealised and unlikely to occur in their idealised forms in practice. Bennett's three configurations (which he terms gestalts) are programmed organisations, professional organisations and problem-solving organisations.

They have ideas in common with Mintzberg's machine, professional and innovative configurations respectively.

Bennett defines his gestalts as follows:

> 'Programmed organisations are so called because their work is highly rationalised. As a consequence, the organisations set up to deal with individual projects are very simple. However, they form part of much larger organisations which take full responsibility for the design, manufacture and construction of standard buildings, bridges or other standard constructions.
>
> The second type of gestalt is professional organisations. They are so called because their work is based on using professional skills and knowledge within the boundaries of established technical rationality. Professional construction project organisations depend on the existence of some form of traditional construction. That is a form of construction where designers know the performance they will achieve from using any particular combination of design details. Also, the local construction industry knows the nature of the work required to manufacture and construct any particular combination of design details. Local contractors know the sequence of specialist contractors required, the effective construction methods, and the plant and equipment needed, and can predict the resultant costs and times with confidence.
>
> The third type of gestalt is problem-solving organisations. These are organisations which produce innovative constructions efficiently. They are called problem-solving because they are set up to find answers to customers' needs which cannot be met by established answers.'

Bennett makes reference to Mintzberg's ideas of organisations being subject to the seven forces as illustrated in his pentagon but believes that these ideas are not relevant to construction projects as project organisations are 'small enough and short term enough to make it possible and appropriate for them to adopt the pattern of a pure gestalt'. It is debatable whether Mintzberg's approach to the synthesis of organisational forms is transferable to construction projects. Mintzberg's work is directed at firms as organisations and the structures are essentially intraorganisational. On the other hand, Bennett is synthesising project organisations which are usually, although not always, a coalition of a large number of independent firms. The issues which manifest in project organisations are about process and different from those within firms. The firms which comprise a project coalition, of which there can be many, include the client, the professional consultancies, the contractor and subcontractors. Each of these companies can be categorised as one of Mintzberg's seven configurations or as a hybrid from his pentagon of five forces. The nature of the project and hence the process will determine which configuration is likely to make a firm the most appropriate for each major task.

The starting point will be the client organisation which could be any of Mintzberg's configurations. Then, for example the design firm is likely to be a 'professional organisation' but may well tend towards an 'innovative organisation' and in fact should have such a configuration for demanding one-off projects requiring unique design solutions. The other professional consultancies could have similar profiles but with engineers and quantity surveyors more likely to be more biased towards a 'professional organisation'. At the construction stage of a conventional project the contractor and subcontractors will tend towards a 'machine organisation' but the nature of construction is that elements of both 'professional' and 'innovative organisations' will probably be necessary. Some of the smaller subcontractors may be mainly 'entrepreneurial organisations'. Depending on the contract strategy, the construction firms may need to be more strongly biased towards 'professional' and 'innovative organisations', for instance in design-and-build and build-operate-transfer projects.

The issue is therefore the nature of the project organisation which overlays and comprises a range of firms each of which has a structural orientation which suits its particular contribution to the project. It should be added that the selection of the firms to be part of the project coalition should ideally be made so that their configuration reflects the needs of the particular project. Straightforward projects would be suitable for professional consultancies with a 'professional organisation' configuration and the contracting firms with a 'mechanic organisation' configuration.

The issue of the distinction between managing firms and managing inter-firm project organisations is one which is not addressed in the mainstream of management literature and will be returned to at the end of this chapter.

Chaos and complexity theory

Chaos theory or chaos science is the forerunner of complexity theory. The nature of the ideas which underpin it do not lend themselves to definition. For example, Gleick's (1993) definition of chaos is 'where chaos begins, classical science stops'. Chaos is used here not in the traditional sense of totally disorganised and random, but in the scientific sense, to describe disordered events in the universe which appear to be random but in fact are not. McMillan (2004) describes this paradox: 'this apparent "chaos" is not an aberration in the planned scheme of things, but reflects deeper more complex patterns and swirls of order than had previously been expected and understood. They are processes that have their own kind of internal order and their own kind of process principles'.

The most well-known example of chaos theory is the 'butterfly effect' which reflects the findings of Lorenz, the pioneer of chaos theory, that a tiny change in a system's initial state does not inevitably lead to small-scale consequences but that minute change can alter long-term behaviour very significantly. The widespread presence of such non-linearity is seen to make prediction impossible over large swathes of the natural and social sciences (Jackson, 2000).

Whereas chaos theory arose from mathematical applications to different fields in the physical sciences, complexity theory is wider ranging and is used to describe the behaviour over time of complex human and social, as well as natural, systems (Jackson 2000).

McMillan highlights the blurring of the divide between complexity theory and systems theory by pointing out that 'some writers refer to the study of complex systems rather than complexity', and cites Allen's (2001) description of a complex system: 'It is any system that has within itself a capacity to respond to its environment in more than one way. This essentially means that it is not a mechanical system with a single trajectory, but has some internal possibilities of choice or response that it can bring into play'. This has similarities with strategic contingency theory.

Many writers recognise that there are neither clear definitions of chaos and complexity nor an accepted chaos and complexity theory. But McMillan offers a working definition of complexity science as being concerned with 'the study of the dynamics of complex adaptive systems which are non-linear, have self-organising attributes and emergent properties'.

In its application to management, complexity theory believes that stable organisations will not generate innovations and will live in the past. Hence organisations should develop a level of instability which will encourage innovation. In particular, a strong shared culture leading to 'groupthink' should not be allowed to develop, as this will lead to a moribund organisation. Rather, different views should be encouraged, as conflict and contradiction will generate creative energy which will produce a learning organisation able to rethink its future.

The spectrum of stability ranges from 'stable' or 'equilibrium' to 'unstable' or 'anarchy'. The 'edge of chaos' is a desirable state that exists 'far from equilibrium'. It is attained when an appropriate degree of tension exists between an organisation's 'legitimate' system and its 'shadow' system. Complexity theorists (Stacey 1993) believe that both extremes should be avoided. In the stable zone they ossify, in the unstable zone they disintegrate but, at the 'edge of chaos', spontaneous processes of self-organisation occur and novel patterns of behaviour can emerge. Jackson (2000) continues: 'The 'edge of chaos' is a paradoxical state where the legitimate system seeks to sustain the status quo and prevent anarchy, while the shadow system tries to change things'. For Stacey (1996), ambiguity abounds; for example, the issue of centralisation versus decentralisation should be ever-present in organisational life, and is best resolved through spontaneous self-organising processes giving rise to new patterns of behaviour. Organisations operating at the edge of chaos deal with such paradoxes the best. It is important, therefore, that managers know how to sustain organisations in this state.

The message for managers is that they have to realise that they should not spend time planning, organising or controlling their organisations but have to create conditions which allow their organisations to be self-organising and learning. As complex adaptive systems, McMillan (2004) believes organisations will learn a great deal about their own dynamics. In order to survive in today's fast-changing world, organisations will need to be constantly innovating, constantly adapting, and continually evolving. McMillan continues by quoting Stacey's (1996) suggestion that:

'organisations operating as healthy complex adaptive systems, do have a special kind of order which they create themselves. There is no mission statement nor a charismatic leader involved in achieving this, it arises spontaneously. These systems are spontaneous, emergent and creative and also very paradoxical. They are paradoxical because they are both competitive and co-operative. Self-organising teams given the freedom to behave as they wish can explore areas normally that would be considered out of bounds as either too contentious, too politically sensitive or whatever. This enables them to explore a wide range of issues and to create and consider innovative and exciting options for handling them. This opens the windows to radical new visions of the future'.

The relationship between systems theory and chaos and complexity theory appears to be contentious. On the one hand, Jackson (2000) finds it ironic that the aspect of systems theory seen to have the most general application today is the study of chaos and complexity, whereas system theory in its origin was about order and regularity. Nevertheless, he sees chaos and complexity theory as important to a systems approach to management and states that 'it is certainly a systems approach, encouraging holistic explanations and eschewing reductionism', an opinion supported by Flood (1999) and Scott (2003). On the other hand, whilst recognising system theory as a significant extension of scientific management, Stacey *et al.* (2000) believe that systems theory excludes 'recognisably human behaviour from the specification of a system and the regulation of explanations of that human behaviour to reasoning processes that are themselves not adequately explained.' System thinkers disagree. But it is the ideas which are important, not the spats between different camps.

The application of chaos and complexity theory to social systems has been subject to serious criticism and can, at best, be considered a work in progress. Jackson (2000) cites arguments not only that the applicability of chaos and complexity to natural systems has not been shown empirically (although it may have been observable in computer simulations) (Rosenhead 1998), but also that human systems are fundamentally different, because humans learn and consequently adapt both their own behaviour and their environment (Johnson & Burton 1994). Rosenhead continues: 'It hardly needs saying that there is no formally validated evidence demonstrating that the complexity theory-based prescriptions for management styles, structure and process do produce the results claimed for them'.

In terms of complexity theory providing an illuminating metaphor for use in management and organisation theory, Jackson goes on to cite Carrizosa *et al.* (1998) who argue that 'the realities the complexity metaphor claims to highlight can be perfectly expressed and tackled using available organisation metaphors, indicating that, in a way, it has nothing new to say. Certainly previous work on informal groups, group working, open systems, emergence, organisations as information-processing systems and turbulent field environments seems to cover much of the territory that complexity theory wants to claim as its own'.

In construction the notion of complexity invariably relates to technology, that is, to project complexity (Fewings 2005). But do the ideas of complex human and social systems have any relevance to construction? Many in the industry would probably say that construction has had its own form of chaos for many years and does not need more! But seriously, although complexity theory may not be a proven theory, the implication of its underpinning ideas for innovation, organisational learning, adaptation and change are worth considering. It is probably the case that their relevance is conceived more clearly for the firms which make up project teams than for the teams themselves, unless all specialist contributors are within one firm, for example a developer with all functions in-house. However, the tensions created by operating at 'the edge of chaos' are alien to the discipline needed and embedded within the culture of firms involved in construction as they seek to deliver tightly defined projects within time and budget constraints. It could be argued that the design firms may more readily relate to these ideas and that they are necessary approaches to the design and construction of innovative buildings. Importantly, there is also the risk inherent in moving to the instability engendered by working at the 'edge of chaos' However, if these ideas are rejected as impracticable, in what ways are firms in the industry going to respond to the need to innovate, learn and change?

The opportunity for project management and project teams to adopt these ideas is even more problematic as the stimulus for innovation, learning and change is inhibited by the structure of the team unless it is all in-house. It is interesting therefore to note that McMillan (2004) gives, as an example of a company which is operating as a complex adaptive system, an advertising and communications company which, in order to move to this state, adopted a project team approach to deliver its services. It should also be pointed out that the company was owned by all its employees, not something one finds often in construction. So perhaps the organisational forms taken for granted by the construction industry are innovative in the eyes of other industries. These ideas could also be useful in understanding the philosophy and organisation of client companies which may have adopted them. The organisation of high-tech companies in areas such as IT, biotechnology and other fields of research may well reflect complexity theory, and this may condition the way in which they deal with project teams in the briefing process.

Postmodernism

A collection of ideas and critiques have emerged which have challenged established assumptions about the way in which organisations are analysed. These ideas reject the idea that there is a reality which can be objectively observed and measured so that universal laws or features or relationships within and between organisations can be established.

Postmodernist conception substitutes interpretation for explanation, as described by Scott (1992):

'An interpretive approach to analysing social systems signifies a number of important changes. First, as Agger (1991) observes, "postmodernism rejects the view that science can be spoken in a singular, universal voice". Rather, "every knowledge is contextualized by its historical and cultural nature". Different truths are associated with differing social or temporal locations. "Social science becomes an accounting of social experience from these multiple perspectives of discourse/practice, rather than a larger cumulative enterprise committed to the inference of general principles of social structure and organisations." A related difference: all knowledge is self-referential or reflexive; that is, we are interpreting a subject's understanding of events in the world not only in their relation to one another, but to ourselves.'

The implications of these ideas for project management in construction, if any, is not likely to emerge for many years but Green (1994) tentatively addressed the nature of social reality and its relevance to interpreting the briefing process for projects and suggested that the dominant paradigm of UK building procurement during the 1980s was based on positivism whereas the dominant paradigm during the 1990s is based on social constructicism.

Green (1998) developed these ideas further by examining business process re-engineering (BPR) from a postmodernist perspective. He concluded that, as many managers in construction seek to influence their organisations through the language of BPR, reality is partially constructed by the metaphors and rhetoric of re-engineering. He sees this in direct opposition to the modernist view that language describes something which already exists 'out there'. As a result he considers that BPR does not possess any universally accepted substantive content other than the rhetoric in which it is presented and is therefore better understood in terms of postmodernist discourse. He continues by extending this argument to say that the reality of the construction industry cannot be understood in isolation from the rhetoric of the management gurus and that the 'problems' of the construction industry do not exist in isolation from the cognitive frames of practising managers whose mental constructs are continually influenced by the popular management gurus. These views undermine the arguments in favour of re-engineering construction as typified by Betts and Wood-Harper (1994) that portray BPR as an innovative, emerging management theory. Not only would postmodernists not accept it as a theory, neither are mainstream organisation theorists likely to grant it such status.

Jackson (2000) is of the opinion that complexity theory 'shares with postmodernism a distrust of rationality as a vehicle for achieving social progress'. He continues: 'Perhaps these ideas have such currency because they offer intellectual succour to the political argument that there is "no alternative" to the market for ordering our social affairs'. But he believes that complexity theorists consider that there is sufficient 'order' for them to be prescriptive to managers about how to improve their performance. He sees the development of interpretive approaches, which are still work in progress, being taken on board by organisations-as-systems approaches in order to attempt to deal with messy, ill-structured problems; but this is by no means universal, and vigorous debate continues.

Critical theory

Traditionally studies of organisations have mainly used models based on rationality, reason and positivism rather than an interpretive approach, as referred to when discussing postmodernism. The traditional approach is challenged by critical theory as well as by postmodernism. Critical theory reflects the views of conflict theorists who see organisations as systems of domination in which groups within organisations (or individuals) seek to take advantage of others. Differences are resolved by the more powerful suppressing the weaker without negotiation or reconciliation (Scott 2003).

Scott believes that 'modernist views of organisations see them as structures for co-ordinating activities in the pursuit of specialised goals. Critical theories assert that such structure can be created only by power used to suppress diverse interests . . . ,' specialised goals being those chosen by the most powerful. Alvesson and Deetz (1996) believe that 'organismic and mechanistic metaphors dominate, thereby leading research away from considering the legitimacy of control and political relations in organisations'.

The ideas embodied in critical theory may be apparent in construction organisations but they can often be more clearly seen within client organisations as conflicting objectives emerge during the briefing process; this is referred to in more detail in Chapter 5 and reflected upon in Chapter 8.

The transaction cost approach

The transaction cost approach emerged from the seminal work of the economist Coase (1937) on the theory of the firm in which he advanced his theory of the existence of firms. He generalised this by stating (Coase 1988) that:

> 'The existence of transaction costs will lead those who wish to trade to engage in practices which bring about a reduction of transaction costs whenever the loss suffered in other ways from the adoption of those practices is less than the transaction costs saved.'

Thus, put in simple, but inelegant terms, those who wish to trade will do so through firms which exist in the market or through creating a new firm to carry on the trade or through expanding their existing firm to cope with the additional trade, depending on which alternative generates the lowest transaction costs. Coase therefore went on to say that, in the absence of transaction costs, there is no economic basis for the existence of the firm and that the limit to the size of the firm is set where its costs of organising a transaction become equal to the cost of carrying it out through the market. Markets and firms are alternative instruments or governance structures for completing a set of contracts.

Coase originally used the terms 'the cost of carrying out a transaction by means of an exchange on the open market' and 'the costs of market transactions' which was ultimately termed 'transaction costs' by others. Dahlman (1979) described them as 'search and information costs, bargaining and decision costs, policing and enforcement costs'. Put very simply in management terms, Coase's approach says that a firm will go to the market for the goods, services, etc., it needs rather than provide them from within the firm when the transaction costs in doing so are lower, and vice versa. It is not an argument that the market should always be the option, which is the way in which his work has often been interpreted by others. What he does say is, 'What my argument does suggest is the need to introduce positive transaction costs explicitly into economic analysis so that we can study the world that exists,' but he believes this has not been the effect of his work and makes a plea for objective evaluations based on his work.

Williamson's (1975, 1979, 1985) work stemmed directly from Coase and gave a particular perspective to the transaction cost approach. It has been Williamson's work which was picked up in the management literature rather than Coase's. Williamson (1979) argues that the difference in governance costs stems from the motivation, cognitive limitations and the moral character of the people involved. His assumptions about the people involved include material self-interest, bounded rationality and opportunism. As a result transactions cannot be achieved unless both parties have confidence in the arrangements which overcome the negative effects of these assumptions. If they are not overcome, transaction costs will rise and may make other organisational forms more attractive. Griesinger (1990) believes that 'if researchers can identify those dimensions of organisational transactions that most affect the costs of governance, it should be possible to choose the most economical organisational form under any circumstances'. Williamson (1979) believes that frequency, uncertainty and asset specificity were the three critical dimensions. Asset specificity means the degree to which the value of an investment is dependent on maintaining the specific relationship in which it is involved or used.

Griesinger (1990) identifies that:

'The transaction cost approach has been applied at three levels: (a) the boundaries of the firm, that is, determining which activities should be governed internally and which should be contracted outside; (b) the overall structure of an enterprise and the relationship of its operating parts, that is, distinguishing between various corporate forms such as the functional, holding company, and multidivisional designs; and (c) the internal organisation of human assets, that is, matching the internal governance processes to the attributes of the workers, their groups, and their tasks. For each case the logic is the same: first, the defining transactions are identified and classified according to the underlying dimensions of frequency, uncertainty, and asset specificity, and next, efficient governance relationships are sought that will protect against opportunism.'

Griesinger and others have traced the exchange perspective to Barnard (1938), Simon (1947) and March and Simon (1958) and to the resource dependency perspective and state that it has become increasingly pervasive throughout social science but he, like many others, does not refer to the fundamental work of Coase. What has emerged is a body of theoretical knowledge labelled by many as organisational economics (OE) (Hesterly *et al.* 1990), in the formulation of which economists have increasingly focused on issues traditionally the remit of organisation academics.

Agency theory is also seen to be an element of organisational economics and is well stated by Donaldson (1990a):

'Agency theory holds that many social relationships can be usefully understood as involving two parties: a principal and an agent. The agent performs certain actions on behalf of the principal, who necessarily must delegate some authority to the agent (Jensen & Meckling 1976). Since the interests of the principal and agent are inclined to diverge, the delegation of authority from the principal to the agent allows a degree of underfulfillment of the wishes of the principal by the agent, which is termed agency loss. Agency theory specifies the mechanisms that will be used to try to minimise agency loss in order to maintain an efficient principal–agent relationship.'

The agency problem is particularly prevalent in construction in cases when clients (principals) do not have the in-house skills to develop their projects so have to hire consultants (agents) and contractors (agents) in the market. The need for clients to protect themselves is caused by asymmetry of information (that is: agents know more than clients) which means that clients need to ensure that agents do not take advantage of them but work entirely on their behalf. Consultant project managers would argue that employing them offers clients the greatest protection, but nevertheless consultant project managers are also agents.

The conjunction of economics and organisational theory has, not surprisingly, caused conflicting views to emerge on the substance of organisational economics. Whilst the transaction costs approach has enriched understanding of organisations and has for the first time established why organisations exist and how they are formed (Hesterly *et al.* 1990), nevertheless the assumptions underlying its more recent applications to organisations (as opposed to Coase's original treatise) has created much debate, not better recounted than in a discussion in 1990 (Donaldson 1990a,b; Barney 1990). Donaldson argued from the perspective of management theory and Barney responded from an organisational economics perspective. Donaldson objects to the assumption that managers behave opportunistically (which is defined as the inclination to lie, cheat, steal, shirk, etc.). Barney argues that this is not a required assumption of organisational economics. He argues that the cost of distinguishing between opportunistic and non-opportunistic behaviour is in fact a transaction cost.

Donaldson asserts that organisational economics adopts a narrow basis of analysis whereas management theory adopts a holistic system level analysis. Barney agrees on the reductionist stance of organisational economics but argues that many management themes are similarly reductionist. Organisational economics' view that managers always act in their own self-interest is refuted by Donaldson. Barney agrees and believes that organisational economics requires the ideas of more sophisticated models of motivation.

Barney does not believe that these differences will prevent an integration between the models but that much of the division is the fear of some traditional management scholars of economic imperialism and that this fear is not unfounded. In a response to Barney, Donaldson (1990b) believes that 'organisational economics should be encouraged as one class of organisational theory but without being allowed to dominate other theoretical approaches and with an eventual aim of integration in a unified organisation theory'.

It seems that the debate, whilst interesting, is also to an extent artificial. If the argument returns to Coase's theory of the firm it follows (Hennart 1994):

'In conclusion, a firm will earn rents if it can reduce its organising costs over those which are shouldered by its competitors. A firm can use three strategies. The first one consists in reducing the cost of organising interactions within the firm by devising and enforcing better employment contracts. A second strategy is to increase the efficiency of external contracts. If a firm can organise co-operation with other firms at lower costs than its rivals, then it can share in the additional gains of trade and garner additional rents. Thirdly, a firm may be more skilled than its competitors at assessing the relative costs of each. It may therefore earn rents by shifting from an external transaction to an internal one, and vice versa, if by so doing it lowers its organising costs.'

The debate continued with power being introduced by Rowlinson and Procter (1997) who view power as equally or more important in explaining the prevalence of the capitalist firm than the economists' view of the firm as an efficient form of organisation, since 'in the real world boundedly rational human actors make socially constructed choices subject to coercion from institutions which enforce agreements'. But there seems to be increasing recognition that integration of transaction cost and organisation theories will lead to a better understanding of organisational design. This is well illustrated by Roberts and Greenwood (1997) in relation to the integration of institutional theory and transaction cost economics from which they build a constrained-efficiency framework. They demonstrate how the admission of cognitive constraints and institutional influences modifies the comparative-efficiency framework of transaction cost economics through recognition of current organisation designs as efficiency seeking rather than efficiency maximising.

Management theory should seek to identify how the various units (internal or external) which generate transaction costs can organise and manage themselves in order to reduce transaction costs and so identify the organisational form which provides the best strategy. In so doing, firms will be managed in a particular manner which encompasses, to a lesser or greater effectiveness, organisation, power, leadership, motivation and all the other elements which contribute to management. What the transaction cost approach contributes is how (as a product of the employment of management theory) the competitiveness of the various organisational forms compares. Organisational solutions will be preferred (and hence, in the long run observed) only if they offer efficiency gains over other alternative arrangements (Hesterly *et al.* 1990). Management theory contributes to the efficiency of organisation and is therefore surely capable of being integrated with organisational economics.

For all these seeming divisions Smith *et al.* (1995) see that issues of co-operation are fundamental to management success and of increasing importance. Informal and formal co-operation occur. The former is concerned with behavioural norms and is traced from the early work of Fayol (1949 trans.) through to Lawrence and Lorsch (1967) and their focus on integrating mechanisms. They see co-operation as formal when characterised by contractual obligations and formal structures of control with a formal hierarchy and rules and regulations, which may provide a perspective for integrating organisational economics and management theory. They also point out that co-operation between organisations had only recently been seen as important. This view will be surprising to people in construction where it has always been essential to the success of projects.

A reconsideration of the human side of organisational economics within the framework of transaction cost analysis, particularly relating to the economic role of co-operation raised by Barnard (1938), is introduced by Griesinger (1990). He believes that co-operation is an interpersonal resource with economic advantages for many organisations and what he terms 'betterment' for most participants, again pointing towards integration of economics and management theories.

Bridge and Tisdell (2004) point out that Williamson (2000) recognises the need for researchers to take a pluralistic position in the development of the theory of the firm and they present an argument for the integration of concepts from the literature on transaction cost economics and the resource-based view. They apply their ideas 'in an attempt to more accurately explain and predict the location of the boundary between the main contractor and subcontractor'.

Whilst Bridge and Tisdell argue specifically for the integration of transaction cost economics and the resource-based view, previous arguments in this section have shown the potential for integration of management theories. It is unreasonable to believe that a single existing theory provides the only explanation. Indeed, they point out that 'Coase (1991) argues that economists have tended to neglect the main activity of a firm, which is running a business. More fundamentally, Coase (1991) believes that a more comprehensive theory would incorporate interrelationships between the costs of transacting and the costs of organising, and that this would *eventually* make his approach in "The Nature of the Firm" operational'. Indeed, they point out that Williamson (1999), in his first critique of transaction cost economics, 'effectively accepts the possibility that the firm may be able to deliver a higher level of effectiveness and/or efficiency, and that technology (comprising the whole suite of the firm's resources) may determine the organisational mode far more frequently than he had previously envisaged'.

Transaction cost applications to construction

There have been an increasing number of applications of the transaction cost approach to the construction industry in recent years; of these the majority have been conference papers, with only a minority being formally published in the academic press, indicating work in progress rather than a mature field.

The earliest formally published work appears to have been by Reve and Levitt (1984), who analysed construction contracts as ways of governing construction transactions but also expanded their analysis into other client–consultants–contractor relationships. The analysis was not quantitative but more in the nature of a general discussion on the issues. Subsequently Winch (1989) made a major step forward. He criticised the work of Reve and Levitt on the grounds that they take the object of analysis as the project rather than the firm. He believed that the project is not an economic entity and does not make resource allocation decisions but that the firms do. He considers that the key question is why construction firms choose to contract for construction services rather than provide these services themselves. Nevertheless, analysis is able to focus on the project as a whole if the object of the analysis is taken as the client (the firm) and the question is why the client contracted for the provision of the project rather than providing it itself. Other analyses with clear objects can then stem from this fundamental analysis.

Winch (1989) applies the transaction cost approach from Williamson's perspective (1975, 1981a, 1981b) to construction by asking the question why construction firms choose to contract for construction services, rather than employ the capacity to provide these services themselves. In his analysis he argues that hierarchy (that is, retaining work in-house rather than subcontracting) would significantly reduce management overheads in the industry, citing project uncertainty, project complexity and post-contract bilateral monopoly as the elements leading to the hierarchy response. He states that in particular, the designer/main contractor and main contractor/specialist subcontractor transaction interfaces could be beneficially governed by hierarchy rather than the market.

He then asks why there has been little shift towards hierarchy in the British construction industry. On the contrary he believes that the trend has been towards greater market governance and he cites management contracting and the subcontracting system as illustrations.

On the other hand, Chau and Walker (1994) found that the transaction cost approach was a powerful tool in explaining the extent of subcontracting in Hong Kong's construction industry. Their work illustrates the basic point that the choice of contracts is not random. Rather, it is predicated on the attempt to minimise transaction costs. The number of contracts, the availability of market information about material costs, time and expenditure in negotiating and drafting contracts, quality assurance, and contract enforcement are the institutional costs that arise in subcontracting. That the parties involved are prepared to pay these costs voluntarily indicate that they are more than offset by the savings in other types of institutional costs under an alternative arrangement.

Between these two papers, little was formally published. However, considerable work has been published more recently. On the boundaries between the client and the main contractor are, for example, Winch's continuing valuable contribution (2001, 2002) and Hillebrand (2000). Work on the boundaries between the main contractor and subcontractor continues, e.g., Lai (2000), Winch (2001, 2002), Constantino *et al.* (2001), Miller *et al.* (2002). Other areas of application include contractor selection (Lingard *et al.* 1998) and risk management (Rahman & Kumaraswamy 2002).

A significant recent work is that of Chang (2006a) who examines the economic characteristics of construction procurement systems mainly in terms of transaction cost economics, resulting in the proposition that in selecting a procurement system, the client will face the trade-off between fastest delivery of a project, high flexibility in accommodating change and single-point responsibility for design and construction (lower transaction costs from measurement responsibilities). The work assumes there are two main sources of variation in transaction costs – opportunistic hold-up and measurement-and-assignment problems. Other traditional categories of transaction costs are downplayed as it is believed that they do not vary with procurement system.

The paper's value in general terms is that it is a rigorous application of analysis using a transaction cost model; and specifically in terms of project management because it shows that the structure of responsibility for design and construction (i.e. project management) is important in the procurement decision. It does however apply transaction cost economics in what Bridge and Tisdell (2004) describe as a 'monistic fashion', but Chang (2006b) had previously argued against their view and they had refuted his (Bridge & Tisdell 2006).

Applications to construction have, understandably, focused on the organisation of firms. Whilst some argue against the project as the object of analysis, others have found it to be valuable (Walker & Chau 1999, Turner 2004). Further understanding can be achieved through more applications to the management of projects. Analysis of projects from the client's perspective may produce transaction costs which argue for a different structure than one produced using the construction firm as the focus, and conclusions may pull in different directions. However, some convergence may occur, as illustrated by the increase in design-and-build. Not only will such analysis focus on transaction cost economising for clients, but it will also encourage pluralism by providing a platform for the integration of organisational theory with transaction cost economics.

The need for pluralism in seeking development of a theory of the firm is equally appropriate in seeking a theory of project management. Morris (1994) believed that such a theory is not possible, but that various organisational theories could be brought to bear as appropriate. Others seek an integrated theory and pluralism would seem to be the way forward. Though Bridge and Tisdell's (2004) attempts at integration were criticised (Chang 2006b), the search for integration of theories continues. However, this search is not helped by advocates of single theory solutions such as those in a recent 'debate' (Koskela & Ballard 2006, Winch 2006). Koskela and Ballard appear to argue for project management theory being essentially a theory of production (and oddly, also argue for it to be integrated into a 'project management theory'). In particular they argue against the contribution of transaction cost economics. Their paper makes scant reference to and recognition of the mature body of work on organisation theory and, as Winch comments, their approach 'is associated with the scientific management approach of people such as Taylor'. Winch mounts a valuable defence of transaction cost economics and refers to the well-recognised point that tools and technique are not synonymous with project management theory. Winch also recognises the contribution of organisation theory and in his criticism states that 'the reference to this literature by lean construction theorists is both very limited, and a misrepresentation'. But even his reference to organisation theory is not extensive; and whilst contingency theory is mentioned, its roots in systems theory are not traced. Nevertheless his recognition of the range of theory relevant to project management organisation shows a need for synergy in further development of a theory of project management.

The integration of transaction cost economics with organisation theory is becoming increasingly accepted as a sound approach to understanding organisational design. Application to project management will be addressed later following the development of project management theory identifying the management activities that give rise to the transaction costs of project organisations.

The major problem of applying the transaction cost approach is the difficulty of measuring/ranking actual transaction costs. Only if empirical work is undertaken can the true potential of the transaction cost framework be realised and arguments such as those above accepted or refuted. However, the operationalisation of the transaction cost approach involves costs which may be extremely difficult to measure accurately particularly if such costs encompass intangible costs such as those associated with motivation and, for example, those incurred in distinguishing between opportunistic and non-opportunistic behaviour. Translating sound theory into useful applications will not be easy due to problems akin to those experienced with cost-benefit analysis.

The expectation that the measurements needed to establish whether organisations should exist as markets or hierarchies can be made is perhaps unrealistic. What determines organisational form is likely to be perceptions of costs in the minds of those who make decisions on the type of organisation to use. Hence many organisational forms can exist, particularly in construction, which are efficient to a lesser or greater extent and which respond to the transaction cost framework to a greater or lesser extent. What the limits are we do not yet know.

Projects, firms and process

Whilst much of contemporary organisational theory stems from systems theory – contingency theory, resource dependency theory, institutional theory and the work of Mintzberg – its focus has been on the processes taking place within firms. It seeks to explain how firms should organise and how they should be managed in order to be effective. It was not until the development of the transaction cost framework and organisational economics that the most fundamental question – why do organisations exist? – was answered. Even more recent has been the recognition by mainstream organisation researchers of interorganisation (firm) co-operation.

Whilst organisation theorists may argue the merits of different approaches, many of the approaches, taken together, provide a basis for analysing construction project organisations. What appears to happen in organisation theory is that the distinction is not drawn between interfirm processes and firms but this is fundamental to the understanding of construction project organisations. Even Winch (1989) when considering the firms which make up project organisations sees Williamson's interpretation of the transaction cost approach as an alternative to the socio-technical analysis of the Tavistock school and to differentiation and integration concepts rather than complementary to them.

The development, design and construction of a construction project is a process which requires a range of diverse skills which is well reflected by Mintzberg (1979):

> 'Every organised human activity – from the making of pots to the placing of a man on the moon – gives rise to two fundamental and opposing requirements: the division of labor into various tasks to be performed, and the co-ordination of these tasks to accomplish the activity. The structure of an organisation can be defined simply as the sum total of the ways in which it divides its labor into distinct tasks and then achieves co-ordination among them.'

A project organisation comprises tasks and co-ordination. On the basis of organisation theory, and simply stated, each of the units which carry out the tasks will arise as an organisational entity each resulting from the application of the transaction cost approach. Each of these tasks may be contained in a separate firm or combined within a firm. Each of these firms will require to be effectively organised for the task it is required to carry out. The nature of the organisation of each firm will be determined by the contingency theory or a derivative. Each of the firms will subscribe to one of, or a combination of, Mintzberg's configurations or result from the effects of his forces. For example, the design firms may be professional/innovative organisations and the construction companies entrepreneurial/machine organisations as they cope with different tasks and environments.

Mainstream organisation theory deals with the above scenario reasonably well but in construction those 'effectively managed firms' will have to co-operate in an interorganisation structure – the project team. This will itself require a structure which is appropriate to the task of producing the project as a whole rather than each part of the project.

Williamson's assumptions about the human component greatly shape transaction cost theory. Those of material self-interest and opportunism sit uncomfortably against the need for high level co-operation and integration of the firms in the project process. Transaction cost theory as derived by Coase is a better basis for explaining the existence of the firms whilst the manner by which they co-operate and integrate is better explained by the contingency theory of organisation and the behavioural and motivational elements of management theory. Without these characteristics the achievement, not only of the world's spectacular structures, but also the many commonplace yet complex buildings and infrastructure projects, would never have been successfully completed. Lansley (1994) remarks that when presented with descriptions of the three types of organisation – market, hierarchy and clan – those with extensive experience of construction often remark that the parties of the typical building project come together through the processes of the market, are expected to operate according to the rule of the hierarchy but, in order to achieve a successful project, have to adopt the characteristics of a clan!

As this book is concerned with the process of management of the project as a whole and not the management of the individual firms except insofar as this affects the management of the process, the rest of the book stems from the application of systems theory, particularly contingency theory, to the construction project management process.

Chapter 4
Systems Thinking and Construction Project Organisation

Introduction

For many years project management has been synonymous with a hard system approach. That is, it has:

'...emphasised quantitative techniques in project planning, scheduling and control. Project-network analysis using PERT or CPM, earned-value measurement, variance analysis, cost-estimating techniques, risk analysis, Monte Carlo simulation, sensitivity analysis, cost modelling, and, lately, expert systems are almost synonymous with "modern" project-management approaches and techniques.' (Yeo 1993)

Yet about the same time, Peter Morris (1994) had cause to comment:

'(Project Management) is widely misperceived as a collection of planning and control techniques rather than as a rich and complex management process. Indeed many of the project management specialists themselves perhaps do not fully recognise the real scope of the discipline.'

The focus on techniques, particularly critical path analysis, is reflected in much of the literature and major books on the subject (Turner 1994). It is remarkable that project managers have taken so long to recognise that, in spite of the many advanced techniques, projects in many fields have still been subject to large cost overruns and delays. The engineering background of many project managers has meant that they instinctively adopt a numerical approach to solving problems even when problems do not respond to such advances. Even in the 1990s project managers were behaving as though soft systems approaches had only recently been discovered although they had been around for many years:

'The soft systems approach is, above all, concerned with human behaviour in organizations, and requires radically different skills in its application: a basic intellect, an ability to see more than one point of view, to think logically, to advocate and to communicate become more important than applying scientific methods, searching for some elusive truth and reducing all problems to rigorous mathematics.' (Daniel 1990)

So it is refreshing to see engineers firmly embracing soft systems ideas at the beginning of the twenty-first century. Blockley and Godfrey (2000) state:

'The important idea here is that all hard systems are understood and managed through soft systems. It is helpful to conceive all hard systems as being embedded in soft ones.'

Morris (1994) goes so far as to rename 'project management' as 'the management of projects' to emphasise that the management of design, technology, political forces, cost-benefit, finance and more which are contained in the latter are substantially more important than techniques.

The hard system approach originated in the operational research field which has had limited application in construction. Project management in construction has therefore to an extent picked up a soft systems approach somewhat more readily than other fields of application of project management and has applied systems ideas to the design of project management organisation structures. Even so the majority of books on project management in construction continue to be technique orientated (e.g. Levy 2000), even the parts dealing with organisation, hence there still remains a hard system ethos due to the engineering background of many in construction. Nevertheless, there is an increasing recognition of the relevance of soft systems theory in the more practical publications on construction project organisation (cf. Blockley & Godfrey, 2000).

The greatest benefit of the application of systems thinking to construction project management is the structuring of organisations in such a way as to achieve the client's objective. Only if this is done in the first instance can hard systems approaches to planning and decision making, and even soft systems approaches to structuring problems, be successful. The use of such techniques and approaches within an inappropriate organisational structure can only mean that their results will be inappropriately applied as the systems objectives and/or structure will be unsound.

Systems theory and its derivatives as applied to the management of business organisations, together with organisational economics, are important to an understanding of how construction projects are organised and managed. Organisational economics help to explain why the firms which undertake the work needed to produce the project are formed with their particular configurations of activities. Systems theory provides a framework for understanding how the process of undertaking the tasks needed to produce the completed project within its environment should best be carried out. That is, it helps to explain how the project management process should be structured and hence how the various firms and other organisational units are integrated into a unified process for the production of the project. Systems theory is also relevant to how the firms and their organisational units organise internally and hence how their response to their task and their environment impacts on the project management process as a whole. It is interesting that general project management research, as opposed to construction project management research, has not taken on board systems theory and transaction cost economics in developing project management theory (cf. Soderlund 2004). This could be due to the idea of managing by projects being a relatively new phenomenon in many industries, but one which has been ever-present in construction.

The application of systems theory to business organisations has focused on the contingency theory of organisation which states that organisation is a function of the task to be carried out and the environment within which it has to be performed. Misinterpretations of this approach, as with the transaction costs approach, are not uncommon, for example Rouleau and Sequin (1995) state:

'In this approach, the representation of the individual is simplified, since both employee and manager are relegated to the background, to be replaced by the profit advantages inherent in environmental constraints, which are principally of an economic nature, and which determine the structural arrangement of the organization.'

Whilst the task and environment determine the structural arrangements, profit advantage is not the sole purpose of organisational arrangements. The purpose of the structural arrangements is organisational effectiveness which may be measured by many criteria other than profit. In business organisations market share, long-term relationships and status can all qualify the profit motive. The main point is that objectives should be clearly defined so that organisational structure can be designed to meet them.

According to Scott (2003), the way in which systems theorists see organisations has moved on from the original concept of structure to a process view. This more interactional conception sees organisations as 'fixed entities having variable attributes' that interact to create diverse outcomes (Emirbayer 1997). The process of relational conception is one in which organisations are seen as 'inseparable from the transactional contexts in which they are embedded'. Weick (1969) was the earliest to shift attention from the 'organisation' to 'organising'.

Scott believes that 'open system imagery does not simply relax the more conventional views of the structural features of organisations: it substitutes process for structure' and that:

'Relational approaches celebrate process over structure, becoming over being. What is being processed varies greatly. In some versions it is symbols and words, in others, relationships or contracts, in still others, assets. But in relational approaches if structures exist it is because they are continually being created and recreated, and if the world has meaning, it is because actors are constructing and reconstructing intentions and accounts and, thereby, their own and others' identities.'

The niceties of such distinctions by organisation theorists will probably mean little to construction project managers, who are unlikely to draw a distinction between structure and process as they are overwhelmingly concerned with process. Construction project management structures are invariably seen as an integral part of process. Nevertheless, it is interesting that this points up the distinctive nature of project management which appears to be more in harmony with developing thinking about the application of systems to organisations than the original applications. Rather than question the validity of applying systems theory to construction project management, these conceptions reinforce its relevance.

A more dramatic alternative view is 'disorganisation', which is argued as a counterpoint to organisation theory's focus on structure by Munro (2003). He argues that 'the rationale for endless change – even when made up unwittingly or on the hoof – seems to be going far beyond the earlier emphasis on processes becoming flexible to markets, and even beyond thinking in terms of a shift from closed to open systems'. It is doubtful whether those involved with construction project management would wish to go down this route.

Systems concepts

It is worthwhile to examine the relevance of the major systems concepts to the construction process to see if they can be used to give a better understanding of construction project management. Embedded in the systems approach are a number of common characteristics of systems which, although couched in systems terminology, can be interpreted in terms of the construction process. The universality of the systems approach is demonstrated by the way in which people from diverse industries have found the concepts acceptable and useful when they have worked them through in their own terms.

The basic distinction between open and closed systems has been dealt with earlier, the response of the open system to its environment being its major distinguishing feature. The original rational models of organisations (e.g. Taylor, Fayol, Weber) and the natural models of, for example, Barnard, Mayo, as defined by Scott (1992), were closed system views as they did not formally incorporate interaction with the environment within their systems.

Traditional management 'theory' had a fixed view of management. It evolved around 'principles', which were held to be universal truths about how sound management should be undertaken. The principles were considered to be the only way to manage business activities or processes, irrespective of the external conditions in which they were carried out. Many of the earlier concepts in the social sciences and in organisation theory were therefore closed system views because they considered the system under study as self-contained. They concentrated only upon the internal operation of the organisation and adopted highly structured approaches. Nevertheless, elements of the rational and natural models continue to be useful in explaining organisation structure and behaviour if incorporated within an open system framework. As a result the open system framework is seen as the most powerful paradigm for integrating other theories of organisation.

Whilst the basic dichotomy of closed and open systems is sufficient for a basic understanding of business organisations it is important to recognise that real-life businesses are to a greater or lesser extent open or closed depending upon the way in which they react to their environment. It is probably only a theoretical possibility that a business system could be entirely closed (it would die) or entirely open. Understanding of systems is aided by Boulding's classification of systems by their level of complexity as summarised by Scott (1992):

'(1) *Frameworks:*   systems comprising static structures, such as the arrangements of atoms in a crystal or the anatomy of an animal.

(2) *Clockworks:*   simple dynamic with predetermined motions, such as the clock and the solar system.

(3) *Cybernetic systems:*   systems capable of self-regulation in terms of some externally prescribed target or criterion, such as a thermostat.

(4) *Open systems:*   systems capable of self-maintenance based on a through-put of resources from its environment, such as a living cell.

(5) *Blueprinted-growth systems:*   systems that reproduce not by duplication but by the production of seeds or eggs containing preprogrammed instructions for development, such as the acorn – oak system or the egg – chicken system.

(6) *Internal-image systems:*   systems capable of a detailed awareness of the environment in which information is received and organised into an image or knowledge structure of the environment as a whole, a level at which animals function.

(7) *Symbol-processing systems:*   systems that possess self-consciousness and so are capable of using language. Humans function at this level.

(8) *Social systems:*   multi-cephalous systems comprising actors functioning at level 7 who share a common social order and culture. Social organisations operate at this level.

(9) *Transcendental systems:*   systems composed of the "absolutes and the inescapable unknowables".' (Boulding 1956)

As Scott points out the nine levels are not mutually exclusive as each higher level system incorporates the features of those below it. It is possible to analyse level 8 using any of the levels lower than 8, 7 using any levels lower than 7 and so on. Boulding believes that 'much valuable information and insights can be obtained by applying lower level systems to higher level subject matter'.

A further basic concept is that of organisations as hierarchies of systems. Hierarchies in this sense does not refer to levels of authority in the classical management meaning but to the arrangement of subsystems, systems and supersystems. Each system is part of a larger system and also comprises other systems. If the physical entity of a building is seen as a system, it can then be conceived as part of the collection of buildings on the road on which it is situated, which is the building's senior system, the buildings on the road are part of the district system, which is a subsystem of the town system which is a subsystem of the country and so on. To understand a system it is necessary to look outside the system at the system in which it is contained which can also be conceptualised as part of its environment depending on where the system boundary is drawn.

A further basic idea is that of loosely coupled systems. The fundamental idea of a system is that it consists of interdependent parts. The impression given is that they are tightly and strongly linked and change in response to each other. This can be seen as an overgeneralisation. It is possible that the parts are relatively weakly connected and capable of autonomous action not requiring a response for other parts to which they are connected. This view is analogous to the fully closed, fully open perceptions of systems' reactions to their environments. The parts of the system may be on a scale from tightly to very loosely coupled but nevertheless they are interdependent. The parts of the project management system tend to be tightly coupled in terms of technical dependency but even this can vary. The degree to which they are organisationally coupled in terms of organisational development, long-range objectives, etc., is certainly variable depending to a large extent on the organisational configuration adopted, as required by the task and the environment of the project.

Objective

A system has an objective. The objective should be stated as clearly as possible and developed as further information becomes available. The manager of an organisation must ensure that all members of the organisation are aiming for the same objective, and must attempt to resolve conflicts where they occur. Many business organisations find it difficult to identify their objectives explicitly but it is an important task of the manager to identify as clearly as possible the objective of the organisation, communicate it to the members and gain their acceptance. If the objective is unacceptable to members, it will be difficult, if not impossible, to avoid conflict, which is damaging to the performance of the organisation.

At first sight the objective of a construction project management system does not seem to be too difficult to visualise and in some ways it is probably easier to establish than for many business organisations. However, the extent to which the interests of a project's stakeholders have to be taken into account complicates the concept and establishment of a project's objectives. The system's objective is typified by the 'client's brief', in which the client states what he expects from the finished project. However, many clients' briefs are unsatisfactory. Often the client's requirements for cost and time for completion are not stated clearly or are incompatible, and sometimes the functional and aesthetic needs of the building are not fully or properly established. This may be caused by uncertainty created by the conditions in the project's environment, in which case the system has to respond by attempting to find ways of coping with uncertainty. It may be through lack of skill or attention in developing the brief, in which case the project management process has been deficient. In either case, it is of paramount importance that the state of development of the objective be known and understood by all the contributors to the project. In the former case, they will be aware of the degree of uncertainty inherent in the objective and should adopt approaches and techniques that can best allow for this. In the latter case they should be allowed to respond to the brief in order that they can contribute to identifying and rectifying deficiencies, so that they work towards an objective which they believe will satisfy the client.

Although the system's objective may be relatively easy to perceive, it may be difficult to articulate. The chances of conflicting objectives arising on construction projects are quite high as a result of most projects being developed by a group of independent firms and professions. The objective of a firm may conflict with the objective of the project team, and the sentience of the different contributors may lead to conflicting interpretations of the project's objectives. The manager of the project will therefore have to set the project's objectives. He must ensure that they are accepted, understood and interpreted consistently by the contributors and must attempt to resolve any conflicts as they arise.

However, systems theory has rather more to say about objectives than this rather simple view. For instance, it considers long-term and shorter-term objectives, the latter often referred to as the goal and the former the objective, but these ideas are often more useful to an analysis of the firm rather than of the project management process. Systems theory also recognises the equifinality of open systems. This means that open systems can reach the same objective from different initial conditions and by a variety of paths. The issue of equifinality continues to be debated. Gresov and Dranzin (1997) state that:

'More recently, the concept of equifinality has come to mean that the final state, or performance of an organisation, can be achieved through multiple different organisational structures even if the contingencies the organisation faces are the same. Equifinality thus implies that strategic choice or flexibility is available to organisation designers when creating organisations to achieve high performance'.

This view reflects strategic contingency theory and is valuable in providing further insights into the application of systems thinking to project management. The project manager can therefore use a variety of inputs in different arrangements in the organisation of construction projects and can transfer these in various ways to achieve a satisfactory output. Thus the project management function is not necessarily to identify a rigid approach to achieving the system's objectives but is to have available a variety of approaches. This view can be extended further. Not only does an open system not adopt a rigid approach; it does not necessarily seek a rigid solution, but has a variety of satisfactory solutions which may meet its objectives.

In terms of construction project management, this concept reminds us that the satisfaction of the objectives of the client does not have to be achieved by the construction of a project. A variety of solutions are open to the client. A client may decide that rather than building it may be possible to take over another company in order to achieve the objective, or alternatively choose to reorganise its own activities to achieve what is required. Even if it is decided that a building is necessary, the project manager can achieve this for the client in a number of ways, each requiring different inputs and achieving the same or different outcomes, all of which may satisfy the client's objectives. If, for example, the project manager satisfies the client by leasing an existing building, the project manager will use different inputs than if a new building were constructed. If the decision is to construct a new building, there are various arrangements that can be used to provide it, e.g. conventional arrangements, design-and-build, etc. This leads to the recognition that the client is part of the project management system or, more constructively, that the construction project management system is temporarily a subsystem of the client's organisation system for the duration of the project.

Bennett (1991) points out that due to equifinality it is impossible, except for the simplest system, to identify the route which optimises the performance of the system and its subsystems. Hence, project managers cannot hope to optimise the performance of construction project organisations. Project managers therefore satisfice, that is they accept a solution which is sufficient to satisfy the minimum criteria for acceptance. He concludes that project managers have a continuing responsibility to search for better ways of doing things but should do so cautiously because systems as complex as construction projects cannot be fully understood and therefore the effect of changes cannot readily be predicted.

Environment

A system's environment consists of all elements outside the system that can affect the system's state, as defined in the previous chapter. This means that environments can be very complex, yet it is not possible to understand an organisation as an open system without a constant study of the forces that impinge upon it.

Organisational theorists have long debated the nature of environments (Scott 1998) and initially drew a distinction between technical (or task) environments and non-technical (often referred to as institutional) environments of organisations. Their focus in the early years was on task environments but subsequently recognised the significance of political systems, human relations, social systems and other similar influences on organisations (Scott 1998). Now it is recognised that this distinction is not valid as many non-technical forces have a direct influence on the technical aspects of the tasks of organisations. The complexity of environments is illustrated by the difficulty of assessing them empirically. Defining them explicitly and identifying the boundary between a system and its environment continues to challenge researchers. For example, should project stakeholders be seen to be part of the client, or in the client's environment? It is possible to analyse project organisations from either perspective. For a clear focus on the client, it is better to conceive stakeholders as being in a client's environment, which is how they are viewed in this book.

In spite of the difficulties faced by researchers, practising project managers still have to relate the project to its environment using their experience and acumen. The manager should not be concerned only with the internal regulation of the system. As the system has to respond to changes in its environment, the project manager must be able to detect and analyse such changes if he is to adapt the internal organisation of the system in response to them. Project managers will be closely involved with issues and problems within the project system but their actions should be orientated to their understanding of the external influences acting upon the project organisation.

Mintzberg (1989) succinctly describes some basic organisational responses to different states of the environment of organisations generally:

- 'The more dynamic an organisation's environment, the more organic its structure.
- The more complex an organisation's environment, the more decentralised its structure.

- The more diversified an organisation's markets, the greater the propensity to split it into market-based units, or divisions, given favorable economies of scale.
- Extreme hostility in its environment drives any organisation to centralise its structure temporarily.'

Understanding these types of responses in a construction setting is essential for project managers.

The process of providing a project is a response to the actions of the environment. The environment acts in two ways upon the process: indirectly upon the activities of the client of an individual project and directly upon the process itself. At its root it is the action of forces in the environment of the client's organisation that triggers the need for construction work. That is to say, the client's organisation has to respond to certain environmental forces to survive, or to take an opportunity to expand, as a result requiring construction work to be undertaken and therefore providing the construction process with work. It may be that new legislation is enacted, which means that a client's present buildings will not conform, e.g. additional requirements for fire safety in hotel buildings; perhaps the client has developed a new production process to compete with his competitors and a new building is needed to house it; a public authority may be required by law to provide a certain new service which requires new buildings. On an international level, internal or external political pressures may mean that a regime has to provide better housing or infrastructure development. In all these examples the need to construct resulted from events outside the client's organisation (or system).

The environment is fundamental not only to triggering the start of the process but also to what takes place within the process of construction. At a strategic level it will determine how the building should be provided. For instance, the state of the property market may have an important effect upon whether a building is leased or a new building is constructed. Such a decision will, of course, also depend upon the process to be housed and whether it requires a new building or can be housed in an existing property. The technology of the process is likely to determine this and will to a large part be dependent upon technological advances in the environment of the client's organisation, for example recently developed materials and machines. Similarly, changes to the proposed building required by the client during design or construction will normally come about in response to environmental forces acting upon the client's organisation.

The environmental forces acting directly on the design and construction process can affect the ability of the process to achieve what the client wants. For example, high economic activity can produce a high level of demand on the construction industry, resulting in shortages of materials, which may delay the project; industrial action can produce labour shortages; high level and uncertainty of inflation can make estimating and cost control difficult, resulting in overspending.

International projects invariably have extremely complex environments. Not only do the environments generated by the countries in which the projects are being constructed affect them, but also the environments of the countries providing the construction team. The effects can be much more pronounced than for locally produced projects for local clients. These influences are reflected in the instability of many developing countries, for example the economic, political and legal environment in China during modernisation. The action of these forces is also often reflected in material shortages in countries that do not have indigenous material availability and lack control over such supplies.

Winch (1989) draws a distinction between the impact of environmental forces on construction firms and those on construction projects. He believes that the effect of economic forces, market complexity and technological change on construction firms are not strong. Rather he believes that environmental uncertainty arises from the project's environment and the way in which projects are awarded. Such environmental uncertainty, he believes, arises from task uncertainty due to the bespoke nature of construction, natural uncertainty such as the weather and geological uncertainty and organisational uncertainty due to temporary project coalitions. He then identifies a separate source of uncertainty which he terms contracting uncertainty. He sees contracting uncertainty as due to estimating not being an exact science and small changes in the tender success rate leading to large changes in levels of turnover. Contracting uncertainty is certainly not a project environmental factor but can be perceived as one in the construction firm's environment which is strongly linked to economic and market forces which Winch does not believe are significant. Construction firms themselves may dispute the view that their environments are not particularly uncertain. It is important to recognise that uncertainty in construction firms' environments and clients' environments impacts strongly on projects to compound the effect of projects' own environments.

Internationally, the impact of environmental forces tends to be much more severe than Winch indicates, presumably for UK projects. Youker (1992) in a review of the results of project monitoring and evaluation on World Bank projects indicates that many of the key problems of implementation lie in the general environment of the project and are not under the direct control of the project manager. He cites the following environmental problems as examples:

- 'A shortage of local counterpart funds (the government treasury does not have the money that it promised to finance local expenditures such as the purchase of land).
- The inability to hire and retain qualified human resources, especially managerial and technical personnel (government personnel policies and procedures do not mesh with the needs of a temporary project).
- The ineffective transfer of technology and difficulty in building and sustaining institutional capacity.
- Difficulty in changing the policy environment, for example in pricing.
- Inadequate accounting, financial–management systems and auditing.
- A shortage of supplies and materials due to overall economic problems.'

If anyone should doubt the impact of both macro and micro environmental forces on projects they can do no better than refer to Morris's (1994) splendid case histories of famous and infamous major projects which, although mainly drawn from the defence and aerospace industries, also include construction.

The perspective taken here is that of environments influencing the structure of organisations; but organisations can also influence their environment, as proposed by the resource dependency model. This effect can be envisaged in the case of monopolies and powerful multinational firms, but less powerful organisations also attempt to influence their environment through, for example, advertising, lobbying by trade representative bodies and strategies for managing stakeholders. Generally, professional practice, construction firms and the construction industry at large do not have the strength or monopolistic power to influence their environment to any great extent. Therefore, the effect of their environment on their organisation is of paramount importance.

Nevertheless, in respect of defining their 'system boundaries', clients and construction firms do have some choice. On the basis of transaction cost economics, they can take 'make or buy' decisions which in effect define the 'technical boundary of the organisation' (Scott 2003). Producing in-house means the activity is within the boundary; outsourcing an activity places it outside the boundary, for example, a client with in-house design and construction capability compared to one which goes to the market for these services, and also the contractors' choice whether or not to subcontract. Scott sees the 'make or buy' dichotomy as an oversimplification as there is a range of alliances available in relations with suppliers, customers and competitors. In the construction industry they are represented by partnering and supply chain management, which are discussed in later chapters.

The tone of this discourse on the environment is that it is possible to know the environment with a degree of understanding, or at least ascribe some probability to one's expectations of environmental effects. An alternative view is that our perception of the environment is constructed from the view we have of it and not something we can objectively discover or unearth (Palmer & Hardy 2000). Yet how this differs from the interpretation of the environment undertaken by businessmen is not clear.

Environmental forces

Scott (2003) classifies environmental forces at the broadest level into technological and institutional forces, with institutional forces encompassing the symbolic, cultural factors affecting organisations and technical encompassing the materialist, resource-based features. He points out that the technical 'reminds us that organisations are production systems' and that 'up to the mid-1970s researchers concentrated primarily on these technical features and their effect on organisational structures'. He further points out that 'the contingency, resource dependence, transaction cost and population ecology theories all privilege the material-resource environment and its effects'. Scott's 1998 edition used the term 'technical' rather than the term 'material-resource' used in the 2003 edition.

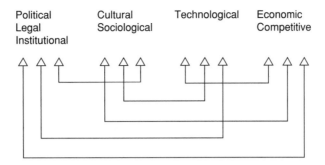

Fig. 4.1   The interdependency of environmental forces.

Now there is clear recognition that institutional forces are probably the more powerful group of forces shaping organisation structures. Forces associated with human relations, politics, law and culture have risen in prominence since the emergence of systems theory. Alongside these developments has been the recognition of the interrelationship of environmental forces. They do not stand alone, but interact in exerting their influence on organisational structures.

To clarify thinking about environments it useful to classify them into general groupings, an example of which is shown in Fig. 4.1. This example could be applicable to any system of organisation, and the forces may be interdependent, as illustrated. It is the interdependency of environmental forces that creates complex environments and makes analysis difficult. As referred to earlier, rather strangely, organisation analysts in drawing a distinction between technical and non-technical forces believed, originally, that they were independent. Members of the construction professions are unlikely to have the same misunderstanding. For example, project teams' knowledge of the rules (e.g. contracts, building and safety regulations) within which technology is applied in construction, and the manner by which political forces influence technological innovation and shape the definition of projects ensure that they are aware of this relationship.

A system receives information, energy and material from its environment, transforms them and returns them as output to the environment. Information is received, for example, regarding the economic climate and the opportunities it presents, new technological advances, the skills of people available to the system and the attitudes of trade unions and employers' associations. Energy is received, for instance, through power to drive machines and provide heat, and through computing power, but perhaps more importantly for the construction system through ideas and people imported into organisations. Material is the raw or partly or fully formed material used by the system, not only building materials but also those consumed by management and administrative processes.

The output of the construction process is returned to the environment. The effect of this can be visualised, for example, as the use to which the client puts the building and the effect on the community of the establishment of the building in a particular location, and, for commercial clients, the effects of enhanced activities on competitors and the economic climate.

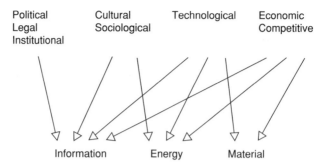

Fig. 4.2   The input of environmental forces.

The forces provide their input to the system in a variety of ways, as shown, for example, in Fig. 4.2. Environmental forces can be classified in a variety of ways and can be identified and analysed for individual projects. From such an analysis the impact of the forces and their input to the system can be anticipated. This approach will give the manager of the project the best chance of coping with them, although it must be recognised that for many projects it is not possible to mitigate the effect of all environmental forces.

The relative importance of the various environmental forces and their impact upon the client's organisation and the process of construction will vary between different classes of client and project. However, the same classes of environmental force will be acting upon each system and can be broadly visualised through the following examples.

Political

By political forces is meant the influence of government policy, for example control of the level of economic activity through investment and taxation policies, and of the distribution of activity through investment incentives, e.g. public–private partnerships. Political forces influence the availability of finance and exert effects on the labour market. Educational and sustainability policies are further examples of political forces. In countries with unstable regimes, international projects are particularly sensitive to political forces.

Legal

Legislation can affect the client's activities by acting directly on the process of construction (for example, through regulations governing building, safety and planning), or by influencing the incentive to build (for example, by controlling the availability of land). In addition, legislation can affect the relationship of participants (for example, through control of monopolistic activity).

Statutory legislation is the result of political activity, but non-statutory law provisions also often have an important effect, particularly overseas. For example,

Islamic law based upon the Koran has an important effect upon how disputes in building contracts are settled in some countries.

Institutional

Institutional forces include the influence of professional institutions upon the activity of their members through rules of conduct, education, and conditions of engagement. Trade and employer associations can exert effects on the activities of their members. Pressure groups, for example the increasing influence of groups promoting sustainability, can have both positive and negative influences on the construction process. The influences of the parent company, head office and shareholders are also institutional forces.

An inverse of these effects overseas is the forces acting upon the contributors to projects in which such institutional constraints do not exist, and their need to respond competitively in such circumstances.

Cultural and sociological

The acceptability of specific activities by the general public, particularly as re-flected by the local community, is an example of a cultural and sociological force. Some stakeholders could fall into this category. The effect of events in the world on the values and expectations of employees is another example of this type of force, as is the influence of trade unions and of informal contacts upon mem-bers of the system. An example of a significant effect of such forces overseas is that, when building within the walls of the city of Mecca, only Muslims can be employed in any capacity on the site.

Technological

Technological forces include the influence of technology on processes through the development of new materials, techniques and ideas and through the experience of others with those materials, techniques and ideas. The current development of technology and its potential for solving problems is an obvious example of a technological force. An extremely important influence in this category is the increasingly rapid development of information technology.

Economic and competitive

Economic and competitive forces include the level of general economic activity and the demands this places upon organisations. The state of competition, the ef-fect of monopolistic phenomena, the availability of finance, materials and labour, and the level of interest rates are other examples of economic and competitive forces. This category is, of course, very closely related to, and dependent upon political forces.

Scott (2003) illustrates (Fig. 4.3) the relative extent to which organisations from a range of activities are influenced by technical and institutional forces,

Institutional Controls

Stronger   Weaker

Utilities Pharmaceuticals Banks General Hospitals	General manufacturing
Construction projects	
Mental health clinics	Restaurants
Schools; legal agencies	Health Clubs
Churches	Child care

Stronger

Technical Controls

Weaker

Fig. 4.3   Technical and institutional forces with illustrative organisations (adapted from Scott 1998)

to which construction has been added. For the purpose of his illustration he uses the terms 'institutional' and 'technical' in a broad sense. He believes that 'most manufacturing and other commercial organisations operate in environments characterised by strong technical and weaker institutional controls' but, as shown by construction having been placed in the stronger/stronger box, construction is increasingly being subjected to extremely strong institutional forces through increasing regulation in such matters as safety, sustainability and environmental controls.

Palmer and Hardy (2000) identify new challenges associated with the current business environment as hyper-competition, enhanced by deregulation and globalisation, short product life cycles, technological change, turbulence, corporate responsibility and customisation. Whilst the construction industry is affected by such environmental forces, the industry's clients are subjected to them to an even greater degree. Consequently projects become increasingly complex, not necessarily technologically but in terms of specifying objectives and hence organisationally.

Other approaches

Two developments of the contingency approach have focused on the effect on open systems of specific environmental components in terms of how they influence organisation structures. Institutional theory focuses on institutional forces in the environment. The interpretation of 'institutional' is broader than that given above. Greening and Gray (1994) state that:

'Institutions specify rules, procedures, and structures for organizations as a condition of giving legitimacy and support (Meyer & Rowan 1977). These institutions have traditionally included state and federal governments and pro-fessional groups as well as interest groups and public opinion (DiMaggio & Powell 1983; Tolbert & Zucker 1983). Also that institutional theory seems particularly well suited to explaining the development of issue management structures because issue management focuses on external influences generated within the public arena.'

Such forces are becoming increasingly relevant to construction, for example in the light of increasing environmental concerns both in terms of pollution and the potential effect of construction work on ecology.

The focus of resource dependency theory emphasises that firms adapt their structures in the face of dependencies on external organisations, and hence man-agers can exercise some choice over organisation structure within the constraints of the environment. The theory believes that organisations are constrained by organisations (institutions) in their environment upon which they depend for critical resources and that hence organisations seek to manage their dependen-cies on external organisations to gain more authority and freedom. It can be seen that resource dependency theory has elements in common with institution theory but that its dependencies are more widely drawn. It also draws from the trans-action cost framework on how organisations are formed. In exercising choice within the constraints of the environment, managers will seek to manage their dependency on external organisations. This is illustrated in construction by con-struction firms' decisions on whether to subcontract or carry out work in-house and whether to own or hire plant.

In developing these theories, elements across all the classes of environmental forces previously identified are incorporated which further emphasises the causal texture of environments (Emery & Trist 1965).

Action of environmental forces

The interaction of environmental forces and their consequent effects on the client and the construction process determine the climate in which the system exists. A low level of activity of environmental forces upon a system will lead to a relatively stable system, whereas a high level of activity will lead to the system existing in an uncertain climate.

In terms of the construction process, environmental forces act in two ways:

(a) upon the client's activities and hence transmitted to the construction process (*indirect*);
(b) directly upon the construction process (*direct*).

The process exists, therefore, in a complex environment, as illustrated in Fig. 4.4, which must be reconciled in the interests of the client. In circumstances where the indirect and direct environmental influences act in a conflicting man-ner, the project management process will be required to attempt to resolve the conflict to the benefit of the client. For example, the contractor may wish to move

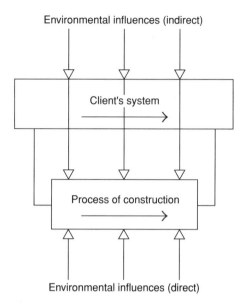

Fig. 4.4   The environment of the process of construction.

labour from the site to aid the profitability of another contract (an influence acting directly on the construction process), which may put at risk the completion
of the building on time when the client's environment demands completion on
time. In such a case it is the duty of the manager of the project to resolve the
issue in favour of the client.

The environmental influences acting directly upon the client's organisation
should determine the organisation structure and mode of operation appropriate to the client's activities. In addition, environmental influences will present
opportunities to the client, and will determine the manner in which such opportunities need to be taken. For example, a client's environment may determine
that an additional manufacturing capacity needs a building quickly in order to
take advantage of an opportunity. In such circumstances, the organisation set
up to achieve this must be capable of acting quickly. If, at the same time, forces
indicate that uncertainty of the size of the market for the goods is likely, the
organisation set up to take advantage of the situation must also be capable of
achieving the flexibility required. However, an unexpected large order for the
client's goods may make the need for a new building urgent, but it may occur at
the time of a rise in activity in the building industry. This may create uncompetitive conditions in terms of price and completion time for projects and make it
difficult to achieve completion when required. Project management must strive
to overcome this type of problem caused by conflicting environments.

The construction process is therefore made complex by the type of environment
in which it exists, which creates a need for high-level managerial skills. The
process must produce a clearly defined solution at the technical level of design and

construction but must also remain flexible and adaptive to satisfy environmental requirements. The managing system will be required to reconcile these competing demands, which become more difficult as environmental complexity increases and in many cases may be incompatible.

The ideas developed here see the process of construction as a subsystem of the client's system. As such, it is influenced by the client's environment as well as by the particular environment of the process. This is a development of the tentative view of the Tavistock Institute (1966) which, although not conceiving the process of construction as a subsystem of the client's system, drew attention to the obsolescent nature of the concept of the architect 'taking a brief from the client' in the conventional way.

Recognition of the construction process as a subsystem of the client's system identifies a boundary between the process and the client's organisation that needs to be integrated. The need for integration has as great an implication for the client as it has for the construction team, since it will demand that both systems establish appropriate ways of achieving the level and style of integration demanded. Originally, a National Economic Development Office (NEDO 1975) report and the Tavistock study referred to the lack of such integration in the conventional process. Subsequent reports focused on the same issue, and the development of project management skills continues to need to concentrate upon this aspect today.

The implication of the relationship between the systems is that changes in the elements of a client's environment or their relative properties may require a change in how the project is designed, constructed or procured. This may happen at any stage in the process. A generic example is the development of partnering, which emerged as a result of client's influence on the project management process. The integrative device at the boundary between the client's system and the construction system should recognise and take action on changes in the client or construction process environments in terms of maximising the benefit or minimising the deficit to the client. This should be the objective of the integrative device.

The relative uncertainty of environments and the nature of the tasks of both the client's system and the construction system should determine the nature of the integrating device and the organisation structure of the construction process. For example, in an environment that is economically or technologically uncertain or both, the organisation structure of the process should be designed to be sufficiently organic to respond to stimuli. This should be reflected in the style of integration used, for example by the project manager assuming a predominant role. This necessity can be visualised, for example, in large-scale long-programme hospital development. Conversely, a stable environment could more readily accept a more mechanistic organisational structure, with integration based on standard procedures and routines, for example in small school building.

Many of the features discussed here are illustrated by Walker and Kalinowski (1994) including the effects of strong macro environmental effects and the manner in which the organisation structure was designed to cope with them.

Whilst the focus here has been on the adaptation of project organisations to their environments, it should also be recognised that projects can influence and change their own environments. The general effect of organisations on environments can be seen in the power wielded by monopolistic multinational companies that exert political as well as commercial influence. At a much simpler level, companies seek to influence their environment through advertising. In between are many other devices used by organisations that seek to control aspects of their environments. On some occasions the impact of projects on the environment can be unanticipated, for example the Millennium Dome has had a significant effect on its political environment changing what was initially seen to be an advantageous involvement with the project to one which was distinctly disadvantageous as the project failed to live up to expectations and harmed the reputation of those involved. The manner in which inappropriately sited projects can distort the pattern of usage of surrounding facilities with ramifications for sociological and cultural aspects is another example: for instance, the effect of out of town shopping developments on city centres; and the power of new airports to influence the location of associated industries and hence related patterns of employment.

An interesting paper by Oliver (1997) on the environment of the Canadian construction industry draws a distinction between institutional and task environments and their effect on performance. She points out that traditionally, the focus has been on task environments, for example markets, resources and competition, but that institutional theory has drawn attention to an organisation's institutional environment represented by state and professional associations' influence in shaping an organisation's legitimacy and performance.

She hypothesizes that an organisation's environment may be strong or weak along either or both task and institutional dimensions but believes that institutional theorists have focused almost exclusively on organisations with strong institutional and weak task environments. The population of her study was the new home-building firms in Toronto so her enquiry concerned firms rather than projects. She chose this population because it was subject to strong pressures from institutional and task environments simultaneously (which supports the earlier arguments about the complexity of environments facing the construction sector). She found that the industry was subject to an elaborate institutional framework to enforce a range of institutional standards and public policy concerns. The regulatory framework included about 280 provincial acts, 460 codes, and over 400 regulations. Up to 25% of the final cost of a project was due to compliance with these regulations.

The conclusions of this serious empirical study found that task environments generally were significantly more critical to organisations' success than institutional relations. Task environmental factors associated with factors of economic production – land, labour, capital and supplies – were more strongly related to profitability and productivity than relations to institutional constituents. Although both categories had some effect, institutional forces were only marginally significant. However, when the institutional environment imposed particularly strong regulatory pressures, the quality of a firm's relations to key constituents of the institutional environment was predicted to have a greater influence on performance than task relations. Conversely, under conditions of high resource stringency, a firm's task environment is a stronger predictor.

The findings show that both institutional and task environments have the potential to contribute positively to performance when environmental forces are severe and recommend that future research should concern the intensity of environmental constraints and the conditions under which institutional versus task environmental relations are likely to predominate in explaining performance.

The author also suggests that institutional theory focuses too narrowly on social endorsement and legitimacy, that institutional constituents also provide strategically useful information on suppliers, competitive characteristics, labour trends, etc. and that high quality relations with the state might enhance an organisation's eligibility for government contracts, grants etc.

Whilst supporting the institutional–task dichotomy for identifying divergent constituents, pressures and constraints, the author believes that the dichotomy tends to deflect attention from environments that impose both institutional and task environmental pressures on organisations simultaneously.

The relationship between task and institutional forces is also reflected in Pries and Janszen's (1995) paper, which examines innovation in the construction industry. They see government regulations and licences having a dominant role but in contrast then find that most innovations were process innovations in the domain of price-cutting, which reflects the influence of the task environment.

Sustainability

A discussion of sustainability is introduced here as an example of how environmental forces emerge, grow and impinge on the project management process and also because it is an important topic in its own right. The increasingly urgent demand for projects to be sustainable has arisen during the last twenty years. It is set to become an increasingly powerful environmental force to which the property and construction industry will have to continue to respond. The groupings of environmental forces referred to earlier have been recognised for many years so it is significant that an entirely distinct force has been created in such a short space of time. It was in 1987 that the Report of the World Commission on Environment and Development was published following four years of study. The Commission's definition of sustainability at that time – 'Development which meets the needs of the present without compromising the ability of future generations to meet their own needs' – is still the most frequently quoted, even though it has been seen as an oxymoron (Lutzkendorf & Lorenz 2005). Nevertheless its message represents the most pressing issue the world faces.

As a result of the continuing depletion of the earth's natural resources, pressure has been exerted on all facets of human activity to protect the natural environment and to conserve resources. That the property and construction industries have had to respond is inevitable and proper in the light of statistics such as those which show that in the USA the construction sector consumes 40% of all extracted materials and accounts for 30% of national energy consumption for its operation, although representing only 8% of gross domestic product (Kilbert *et al.* 2000). The concept underlying the above definition has been operationalised by the so-called 'triple bottom line' of sustainable development, which states that sustainable development involves balancing economic and social development with environmental protection (Lutzkendorf & Lorenz 2005). They point out that there are discrepancies in the interpretation of sustainable development, as the three dimensions can be considered equal or the environment can be viewed as the dominant dimension that sets the preconditions for the others.

As we have seen, environmental forces act on the property and construction industries at large and on clients, from where they are transmitted to the construction project process, and directly onto the process itself. This scenario applies equally to the sustainability environmental force. At the industry-wide level, Kilbert *et al.* (2000) describe 'a view of the construction industry based on natural systems and industrial ecology for the purpose of discovering how to shift the construction industry and its supporting materials industries onto a path much closer to the ideals of sustainability'. At the opposite end of the spectrum of abstraction, Myers (2005) finds that, on the basis of public disclosures made by construction companies listed on the UK Stock Exchange, very few companies embrace the agenda for sustainable development, and that as these companies represent the more sophisticated class of companies, the many small companies are more unlikely to have bought into sustainability. Clients are seen to be the drivers in both the public and private sectors which, together with regulation (an institutional environmental force), will produce a reaction to the sustainability environmental force, but only if they themselves are required to respond in order to survive or develop (the position of clients is pursued further in Chapter 5).

The emergence over recent years of a powerful environmental force with a great impact on the property and construction milieu, which will grow immeasurably in the next few years, reinforces the importance of all classes of environmental forces in the analysis and understanding of how the construction project management process operates.

Measurement of environmental uncertainty

Whilst the ideas of environmental uncertainty can be readily conceptualised, its formal measurement has received little attention by researchers (Buchko 1994) no doubt due to the sheer complexity of environments and hence the difficulty of devising convincing methodologies. Buchko provides a good review of research in this area which begins with Lawrence and Lorsch's original work on uncertainty associated with a specific job in an organisation. Researchers moved this on to applications to the firm but were faced with increasing complexity.

Buchko believes that:

'The conceptualizations of the environmental uncertainty construct used are not consistent, ranging from predictability through dynamism and complexity to controllability. Such differences in the underlying conceptual definitions of the environmental uncertainty construct make generalizations or interpretation of results difficult.

Several studies have used measures of perceived environmental uncertainty that are idiosyncratic to the particular research effort and that their conceptual definitions of uncertainty have varied significantly; such concepts as turbulence, complexity, predictability, and heterogeneity have all been included, making interpretation of results across studies problematic.'

As a result of his own research using the Miles and Snow Perceived Environmental Uncertainty Scale (Miles & Snow 1978) Buchko concluded that:

'Perceptions of uncertainty may be inherently unstable because environmental complexity and dynamism may prevent individuals in organizations from developing stable assessments of the environment itself. Organizational and individual characteristics may affect the perceptual process . . . yet the difficulties posed by the natures of organizational environments and managerial perceptual processes may make the development of reliable and valid measures very difficult.'

Hughes (1989) used a much less ambitious approach in his attempt to describe rather than measure the environments of construction projects. Using eleven environmental factors, e.g. economic, cultural, etc., he scored each on a scale of variability of 0, 1 and 2 for each of five variables – definition, stability, certainty, simplicity and mitigability. A problem with this approach is that the variables may not be mutually exclusive. Nevertheless the approach is useful for helping managers understand the environments of their projects, and for giving discipline to project scanning. Project scanning is the process of analysing the project environment to define potential problems and assessing the probability of their occurrence. Youker (1992) believes that there are several different ways of scanning the environment each of which contain the categories of environmental forces identified previously:

- 'Elements that are suppliers of inputs, consumers of outputs, competitors, and regulators.
- Elements that are physical (e.g. climate), infrastructural (e.g. power supply), technological (e.g. plant genetics), commercial/financial/economic (e.g. banks), psychological/sociocultural (e.g. attitudes toward credit risk), or political/legal (e.g. local government).
- Elements that are hierarchical and sometimes geographical, such as government at various levels, i.e. national, regional and local.
- Elements that are actors (e.g. individuals, groups, institutions) or factors (e.g. attitudes, trends, laws).'

Whilst these illustrations are drawn from World Bank projects, many will be familiar to those involved in construction projects. Youker points out, very practically, that analysis of the key elements of the environment will not solve all problems as some will be intractable but at a minimum will give early warning of problems so that the best chance of finding a solution will be provided.

There is therefore no easy or precise method of quantitatively assessing environmental forces and their impact on construction projects. Indeed, if there were, it would be a large step towards solving the problems faced by governments, let alone by construction teams! Nevertheless, recognition of environmental forces and an understanding of how they affect construction projects does allow those involved in their management to construct a scenario for their work which should allow them to anticipate and respond to changes in environmental conditions. Preconceived methods of working all too often have fixed rigid patterns for project achievement. This can mean that the process is unable or unwilling to respond to environmental changes to such an extent that although a client's objective may have changed, the objective of the construction process has not.

A construction project organisation should therefore be designed to reflect the type of environment in which it is to work. There is little point in using a well tried rigid organisation structure suitable for building public housing for the design and construction of a second Sydney Opera House. The conventional UK approach to organisation of the construction process is not directly transportable to overseas projects for similar reasons. The environmental conditions in which such projects are designed and built are very different from those in which the convention was developed.

Organisational competence

The concept of organisational competence is raised by Olmstead (2002), who sees it as the ability of an organisation to sense the impact of events in its environment on the organisation, process the information and adapt the organisation to cope with the changes. Organisational competence is seen as the determinant of organisational effectiveness.

The crux of organisational competence is seen to be flexibility, and the point is made that many organisations are so hidebound by plans and procedures that they cannot adapt to changing environmental conditions. Olmstead states that the emphasis on 'organisational responses to problem situations points up the role of the organisation as problem solver, decision maker, and action taker'.

The principal device for maintaining this effort is seen as 'the chain of authority which runs through the heart of the organisation from the topmost level to the lowest point of oversight. The various levels in the chain of authority, together with staff roles designated to assist various positions, strive to control and co-ordinate information, decisions and actions so that unified action results'. Whilst reference is made to managers undertaking these tasks, the emphasis on the organisation having competence can be seen to be a simplification which corresponds with Fincham and Rhodes' (2005) identification of organisational goals as oversimplified due to its tendency of attributing human motives to social institutions (known as reification).

Their point is that only people in organisations can have goals (which may conflict), and that it is all too easy to refer to the organisation as the active party so that the people making decisions and the motives behind their actions tend to be concealed. The corollary is that, rather than organisational competence, the competence and characteristics of the people involved in the decisions necessary to adapt the organisation to its environment are the determinant of the organisation's flexibility. Nevertheless, the idea of competence is valuable and not one which receives the attention it merits.

Negative entropy, adaption and protected environments

Entropy is applicable to all closed physical systems. It is the tendency for any closed system towards a chaotic or disorganised state in which there is no potential for further work (Katz & Kahn 1978). This may sound rather dramatic for a book about construction project organisation but it has some significance. Construction project organisations are open systems and open systems attempt to find ways of avoiding such a fate. They develop negative entropy (negentropy), which is a process, of more complete organisation and greater ability to transform resources. They achieve this by importing the resources (material, energy, information) from the system's environment. Social organisations such as those involved in construction can continue to import new human and other resources to allow them to continue functioning and may be capable of indefinitely offsetting the entropic process in a way which open biological systems and closed physical systems cannot.

In developing negentropy, an open system may be seeking to achieve a steady state in which the system remains in dynamic equilibrium through the import of resources from its environment. That is, it attains stability or is self-regulating. This view is more relevant to biological systems and allows them to cope with variations in their environment. For example, the human body can maintain a steady state in spite of wide variations in the environment. There are of course limits; environmental changes may be so great that the system dies.

Taking these ideas a stage further introduces the adaptive system (Buckley 1968). A system is adaptive when it changes its own state and/or its environment when there is a change in its environment and/or in its internal state that reduces its efficiency. Adaptation is therefore the ability of a system to modify itself or its environment when either has changed to the system's disadvantage. Complex adaptive open systems allow interchanges among their internal components (or subsystems) in response to environmental forces to such an extent that the components themselves may change and hence the system as a whole may adapt and so survive.

It is interesting to assess whether the construction process fits into this systems scenario and whether it helps our understanding. The firms that contribute to the construction process do import material from their environments in the form of new staff, new ideas, new technology, etc., and so develop negative entropy. However, construction project organisations are generally temporary. They cease on completion of the project and further organisations will be formed with either the same amalgam of contributors or with different partners. The process itself is not therefore truly negentropic except in the case of design-and-build or turnkey organisations. The firms themselves will, however, be importing from their environments so that they are in a position to be able to join future project organisations. In the case of design-and-build and turnkey organisations, one firm is responsible for practically the whole process and in these cases the firm and the process attempt to develop negentropy.

In adapting to their environment, some systems will attempt to cope with external forces by acquiring control over them. This process can be seen in the mergers of companies, often to reduce competitiveness in their environments, which result in the expansion of the original system. Some organisations may have achieved such a degree of monopoly or have acquired a protected niche in the environment to the extent that they can ignore a certain level of environmental pressure. If this occurs, such organisations can afford to accept a level of suboptimal performance and can survive at that level.

In the case of the construction industry's professional and industrial firms, for many years the amount of adaptation to their environment was not great, as illustrated by the large proportion of projects undertaken predominantly on the traditional pattern in spite of much criticism of this process. The traditional pattern of organisation tended to be self-regulatory and to function to maintain the given structure of the system. This was due, to a large extent, to the system existing in an environment from which it had protected itself. This was achieved through codes of conduct and fee scales of its professional institutions, which eliminated, to a large extent, competition between firms, thus enabling the system to resist change and maintain the status quo.

However, the increasingly competitive environment in which the system and its clients now exist has been significant in breaking down such protection. In the case of clients they have brought to bear greater pressure for change in the industry's procedures as a result of the increased competition with which they themselves have been faced. The increasingly multinational nature of the industry's clients and globalisation have also been major forces for change as clients have experienced methods of managing the construction process that differ from those used conventionally in the UK.

As the professions' and industry's firms are no longer able to protect themselves at an institutional level, changes have taken place at the project level with the increasing use of such processes as design-and-build, partnering, prime contracting and build-operate-transfer. At the project level it appears that some firms are adapting by changing the nature of the internal components of the building process, for example by introducing the contractor into the design team, and so moving nearer to the open adaptive system described earlier.

In view of the traditional institutional domination of the professions' and industry's firms, it was always likely that adaption would take place at the project level in response to the demands of clients. However, this was only likely to take place for projects with clients who themselves were adaptive and not protected in some way from their own environments. It is not surprising, therefore, that many initiatives that have taken place in the management of construction projects have been for private commercial or industrial clients, and that there was a lack of stimulus from public clients who were themselves protected to a large degree from their environments. However, with the change in local government culture in the UK changes also took place in this sector's projects. The process of providing a construction project should be an open adaptive system but it may be constrained by the environment within which it exists. Nevertheless, the process needs to change its structure if environmental events, acting either directly upon the process or indirectly upon the client's organisation, dictate that this should happen.

In taking these ideas a stage further, organisation theorists have identified buffering and bridging strategies by which organisations relate to their task environments (Scott 1998). Whilst buffering strategies are designed to protect supplies and control demand and may not have great relevance for construction project organisations, bridging strategies will be far more readily recognised. They are intended to form a bridge between organisations and thereby enhance the security of their task environment. They include alliances, joint ventures and mergers amongst others (Pfeffer & Salanick 1978) and hierarchical contracts (Stinchcombe 1985). In construction, they would also include the wide variety of arrangements that go under the title of partnering. Bridging can be seen as a response to the recognition of organisational interdependence and uncertainty, which a measure of integration is intended to help to solve.

Growth, differentiation, interdependency and integration

Whereas closed systems move towards disorganisation (Kast & Rosenzweig 1985), open systems move in the opposite direction towards a higher level of organisation, which generates greater differentiation among their parts (sub-systems). This feature is observable in business organisation systems and can occur in two ways. One, which has been referred to previously, is when a system seeks to encompass parts of its environment and annexes them as subsystems, for instance when one firm acquires control of another. The second way by which it occurs is where complex and uncertain environments create the need for subsystems to specialise further in order to cope with such complexity and uncertainty. That is, the level of skill required is such that a subsystem cannot cope with the range of skills demanded of it and it has to subdivide further. In traditional management thinking, specialisation was considered to be 'a good thing' for increasing the efficiency of undertaking a particular task. Differentia-tion, however, is now explicitly considered to be necessary in order to allow each subsystem to cope effectively with the part of the system's environment which is acting upon it. Hence, open business organisation systems tend to grow by expansion and by internal elaboration. This is not to say that this is necessarily a benefit in all cases but simply that it is a feature of open systems. It brings along with it the greater management problems of handling large complex sys-tems and hence the need for careful organisation design if such systems are to be effective.

The protected environment of the construction process originally limited the impact of these concepts upon it. However, there are now numerous examples of growth through amalgamation of professional firms, creation of design-and-build companies and client organisations acquiring an in-house capacity for designing and constructing projects by taking designers and construction workers on to their payroll. Similarly, a number of consortium firms including architects, quantity surveyors and engineers have been formed. All of these arrangements helped organisations to handle more easily the environment in which they operated. A growing area of such activity is the joint-venture movement in which firms form separate joint companies to bid for and undertake projects. Companies make such ad hoc arrangements between different types of firm, e.g. architects and contractors, and between similar types of firm, e.g. contractors. In most cases the aim is to spread the risk in the difficult and uncertain environmental conditions that usually accompany large and complex projects, which are often overseas. Such activities are also predicted by the transaction costs framework, which indicates that management theories are not mutually exclusive.

Growth through internal elaboration has occurred relatively slowly in the construction industry. After many relatively stable years, specialisation into architect, quantity surveyor, specialist engineers and contractors took place quite quickly during the eighteenth century and the early 1900s. This resulted from the increasing complexity of the environment in which construction took place during industrialisation. The process then slowed down as the professions protected themselves from their environment and attempted to maintain the status quo. The subsequent proliferation of specialist subcontractors can be seen to be further differentiation to cope with complexity and uncertainty. There is also evidence of the same phenomenon in the specialisation of quantity surveyors into tender document production, construction economics and cost control. Similarly, architects often specialise in particular building types. Once again such movements reflect the transaction cost framework as well as systems theory in a manner which is the converse to that described above.

The notion of interdependency is explicit in the earlier definition of a system, that is, it is an entity consisting of interdependent parts. If this is the case, then the greater the differentiation of the interdependent parts of a system, the greater will be the need for integration. Differentiation in organisations has been defined (Dalton *et al.* 1970) as the differences in cognitive and emotional orientations among managers in different functional departments and the differences in formal structure among these departments. Integration has been defined (Dalton *et al.* 1970) as the quality of the state of collaboration that exists among departments that are required to achieve unity of effort by the environment.

Scott (1998) makes the point, when considering the technical core as a source of structural complexity of organisational systems in general, that the common element of complexity, uncertainty and interdependence is that each increases the amount of information which needs to be processed during the carrying out of a task. Thus modifications need to be made to the organisation structure that reduce the need to process information or by lowering performance standards or by increasing the capacity of the information processing system, all of which have been proved to be extremely difficult to achieve on construction projects.

The interdependency of the contributors to the construction process has long been recognised but often as sequential interdependency. In other words, one part cannot act until after the previous part has done its work. It has become increasingly recognised that in fact interdependency should be reciprocal, that is, iterative and the process should move forwards following decisions to which all appropriate parts of the system have made a contribution. The integration of reciprocal interdependencies requires considerably more skill and effort than the integration of sequential interdependencies. It is the recognition of this fact that has focused attention upon the management needs of projects. Recognition of the need for project management has been highlighted by the complexity and uncertainty of the environment within which construction takes place, which has led to greater differentiation within the construction process and hence to a greater need for skill and effort in integration.

This has led to many initiatives to increase integration in practice in all aspects of the construction process such as partnering, supply chain management, clustering and increasing use of design-and-build, each of which are dealt with in later chapters. Recognition of the need for integration within a framework of systems thinking has received greater attention, even in the more practice orientated publications on project management (cf. Blockley & Godfrey 2000).

Feedback

The concept of feedback is fundamental to understanding how a system is maintained and therefore how it continues to exist and accomplish its purpose. Feedback is the basis of a system's control function. It is through feedback and subsequent action that achieved outcome can be compared with desired outcome so that adjustments in the behaviour of the system can be made. The need for a control function for construction projects is self-evident and much of the energy expended in developing techniques in recent years has been directed at achieving more sophisticated control. However, the type and the amount of feedback designed into a system are the key to the system's stability and economy and in this respect it is interesting to note that the control mechanisms on construction projects are often no more than monitoring devices that declare the position too late after the event to take corrective action, e.g. many 'cost control' procedures. Feedback points should be carefully designed into the system so that appropriate action can be taken at the right time. Feedback should operate on a cost-effective basis in such a way that the value of the control achieved is not cancelled out by the cost of achieving it.

The operation of a feedback loop requires a sample taken at specifically designed points of the system's operation (often referred to as freeze points) to be measured against the objective of the system. For construction projects the sample points need to be chosen on the basis of the nature of the actual project and its environment. That is, for a simple project in a stable environment it is to be expected that only a small number of sample points will be necessary, whereas for a complex project in an uncertain environment, frequent sampling will be required. Naturally, this means that the objective of the system should be appropriately, accurately and explicitly defined to enable the control mechanism to carry out its function. It is questionable whether many of the 'client's briefs' commonly used in the construction industry are sufficiently clear to allow this to happen. Effective control systems require that the procedure for testing the sample against the objective be designed with appropriate methods of measurement of the sample against the objective and, importantly, with the ability to take action on the basis of feedback information.

The traditional organisational structure of the construction process often does not possess this ability as the relationships of the contributors to the process are arranged in such a way that the people reporting on the current state of the project vis à-vis its objective are not in a position of sufficient authority to ensure that the project returns to its intended course. This is often the case when the architect is in both an operational capacity as a designer and also in the primary management position for the project. The application of a systems approach to the design of organisations should automatically establish relationships which would allow a properly designed control function with appropriate feedback mechanisms to overcome this deficiency and operate effectively.

The simplest kind of feedback is negative feedback. This enables the control function to correct the system's deviation from its course, that is, it encourages a return to the initial objective. Most control functions used on construction projects operate in this way by attempting to correct deviation in cost, time or design of the project and return it to what was intended by the 'client's brief'.

Positive feedback, on the other hand, further amplifies deviation from course, as a result of redefinition of the system's objectives. Although this may be an unusual reaction for construction projects, it should not be overlooked. If, during design, it is discovered that the provision of facilities in the project is deviating from what was intended, it may be that the client's original requirements have changed, and upon seeing the developing design the client may decide to continue along this course. The objectives would therefore require to be amended in response to positive feedback. Such change could be as a result of a badly constructed brief but could also be caused by changes in the environment of the client's organisation that have altered their objectives. The control function should therefore operate within the system and between the system and its environment.

The nature of the process of designing and constructing is characterised by a series of 'pinch points' through which it must pass if progress is to be made. At each pinch point a decision has to be made, for example whether the design satisfies the function required of it, whether the cost is acceptable, whether the proposed procurement method will allow the project to be completed on time, etc. The decision points can be conceived as a hierarchy with decisions taken by the client at the top, those taken by the manager of the project at the next level, and those taken by the operational people at the lowest level. The decision structure of a project can be used to provide the control framework. Feedback can take place at each decision point to test whether the proposed decision will help to achieve the objective of the total system. It should be noted that, on many projects, decisions are not made explicit and therefore are not used in this manner. They are not consciously tested but are assumed to be correct as 'that is the way we have always done it'. However, by anticipating decision points and the nature of the decisions to be taken, a control framework can be established and the contribution to be made by each participant can be designed using systems principles.

Systems and hierarchies

Hierarchies in organisation structures are commonly associated with the classical approach to management and are seen as rigid and lacking the flexibility and adaptability required by today's fast-changing business environment. But is this the case? Leavitt (2005) makes the point that hierarchies are still the reality of the vast majority of large organisations, and asks: Why are hierarchies still with us? Why do we seem to continue to need hierarchical appendages such as rank, status, organisation charts, etc.? It remains an interesting paradox that there appears to be universal dislike of hierarchies, yet we keep recreating them in spite of modern information technology and associated developments which were expected to see their demise. Or is it that they are simply seen to be old-fashioned by management gurus, some scholars, consultants and executives, whilst many that work within them find them quite comfortable?

Leavitt argues that hierarchies provide opportunities to achieve power, status, wealth and social recognition, and hence security. He also believes that they are great devices for coping with complexity, particularly as organisations grow and age. In seeking to explain the continuance of hierarchies, he simplifies organisations by envisaging them in two ways: humanising and systemising. His view of systems is somewhat slanted toward hard systems views, which he believes leads systemisers to favour hierarchies. Humanisers are seen to be opposed to hierarchies, as they focus on the emotional needs of people.

A soft systems view is seen as a better characterisation of approaches to project management organisations, but this does not deny the need for and significance of hierarchy within project organisations. In abstract terms, systems theory recognises the existence of a hierarchy of systems. It follows therefore that in a model of the construction project process that consists of a number of differentiated yet integrated subsystems, there exists the need for an overarching hierarchy. One would expect that an effective hierarchy in construction projects would operate with a light touch, with great delegation to the project management system. It is unrealistic to believe that this would be unnecessary. Put simply and broadly, hierarchies would comprise the manager at the top level within the client organisations, the project manager next, followed by the managers of the professional firms.

Leavitt also makes the point that hierarchical organisations are different from each other, implying that they adapt to their different environments. This can be seen as their behaving as part of an open system. He goes on to say that some large hierarchies introduce matrices and other horizontally and diagonally linked substructures to help cope with increasing, multifaceted complexity and that hierarchies have been flattened, networked and teamed. This reinforces the realisation that hierarchies are in fact part of the system, and that systems theory and the hierarchical structure element of classical organisation theory are compatible even if the behavioural connotations of classical organisation theory are not.

Leavitt concludes his argument by stating:

'Hierarchies remain the most workable and effective structures humans have yet invented for performing large, enduring, complicated tasks. It would be difficult indeed to manufacture and distribute large numbers of tractors, televisions or telescopes without hierarchies, just as it would be difficult for a community of any size to govern itself hierarchy-free.'

The same can be said for construction projects.

Associated with these ideas are those by Blau (1970), as discussed by Scott (2003), on how the size of an organisation relates to its degree of bureaucracy. Contradictory findings emerged because hierarchies are made up of different types of employee who act as administrators – managerial, professional, clerical, etc. Depending on the organisation's task, as organisations grow they may increase the number of specialist units they require, which creates greater differentiation which then requires more administrators (with a range of backgrounds) to provide integration; other organisations may grow due to more, or expanded, activities of the same type, which does not increase differentiation, and which does not require more (specialist) administrators for integration. Interestingly, these ideas on hierarchy reflect the systems idea of differentiation and are relevant to understanding the structure of client organisations.

Summary

Jackson's (2000) reference to Katz and Kahn's (1978) ten characteristics of all open systems provides a succinct summary of an open system:

- 'The importation of energy from the external environment
- The throughput and transformation of the input in the system
- The output which is exported to the environment
- Systems as cycles of events: the output furnishes new sources of energy for the inputs, so the cycle can start again
- Negative entropy: open systems live off their environments, acquiring more energy than they spend
- Information input, negative feedback and a coding process: systems selectively gather information about their own environments and also about their own activities (so they can take corrective action)
- The steady state and dynamic homeostasis: despite continuous inflow and export of energy, the character of the system remains the same

- Differentiation: open systems move in the direction of differentiation and structure elaboration (e.g., greater specialisation of functions)
- Integration and co-ordination to ensure unified functioning
- Equifinality.'

The process of designing and constructing a project on behalf of a client can be analysed as an open adaptive system using these characteristics.

Subsequently, the term 'project governance' has been used to refer to co-ordination and control of the process of managing construction projects. The term has been used to describe the governance of transactions (Winch 2001) and to describe the governance of construction coalitions (Pryke 2005). Quite what the distinction is between the meaning of project governance and project management (drawn in its widest interpretation) is not clear.

As an open adaptive system, the process needs to respond to its environment but historically it has, to a degree, protected itself from its environment by the construction of rules, procedures and conventions which have been granted validity by public authorities, professional institutions and other bodies associated with construction. Nevertheless, the environmental influences upon the process, particularly those being transmitted to it through its clients, have resulted in the process becoming more responsive.

Analysis in systems terms focuses attention upon the need to bind together the differentiated yet interdependent contributors to the process. This requires a high level of integrative activity which has not traditionally been recognised and provided. The provision of integration must be directed towards the achievement of the total systems objective, which must be stated unambiguously in terms to be the client's requirements.

The determining factors of how the system is structured and operates to achieve its objectives are the technical demands of the project, together with the environment in which it is undertaken. The control function should therefore be designed to reflect these factors and be based upon the anticipated decision points in the process. The decision points will determine the interdependency of the contributors to each decision. Therefore their relationships should be designed on systems principles in terms of their contribution to each decision. Such an analysis demands that the organisational structure established for each project should be developed individually from first principles, and although a range of 'standard solutions' may emerge, it should not be presupposed that any one solution is automatically the correct answer.

Returning, briefly, to the environment and its implications for the construction industry and its professions, it has been observed that industrial companies often fail to obtain adequate information about changes in environmental forces and that it is remarkable how weak they are in their market research departments when they are so dependent upon their market. It has been predicted that organisations will improve their facilities for assessing environmental forces and hence improve their marketing ability. The same could be said of the construction industry and its professions now, and perhaps the same prediction can be made.

The remainder of this book is developed against the background of this systems scenario of the construction process, which can give structure to our understanding of the process. The systems view clarifies how the process works in practice at the present time and points the way to how it may improve in the future to enable it to satisfy more effectively the requirements of its clients.

Chapter 5
The Client

Introduction

On the face of it the use of the term 'the client' in the construction industry is simple enough but its apparent simplicity hides a complex concept. As early as the mid-1960s the Tavistock Institute (1966) was drawing attention to the increasingly intricate nature of client organisations, saying:

'that they were complex systems of differing interests and that their relationship is seldom with a single member of the building industry ... These client systems are made up of both congruent and competing sets of understandings, values and objectives. Much design and even building work has proved to be abortive because unresolved or unrecognised conflicts of interests or objectives within the client system have only come to light after the building process has been initiated'.

Since that time clients have been putting their views more forcefully. For example by Mobbs (1976) of Slough Estates who accused the UK construction industry of poor performance and subsequently of failing to satisfy the needs of clients, and particularly purposefully by the British Property Federation (1983) which devised their own system for commissioning design and construction due to their dissatisfaction with conventional methods at that time. It was in response to such pressure from clients that the government embarked upon the series of reports and initiatives beginning in the 1980s running through to the new century, as discussed in Chapter 2, as a result of which the professions and industry responded to clients' criticisms. The outcome has been strong support from experienced clients for the initiatives of the Strategic Forum for Construction (SFfC, 2006). Clients have, therefore, had a great influence on the industry and have induced changes in attitude which have seen the industry adopting a wider range of organisation forms and greater flexibility in seeking to meet clients' needs. At the same time as these changes were taking place, the concept of 'the client' was changing. The concept of the unitary client was seen to be too simplistic in many cases, particularly for large complex projects subject to turbulent environmental forces. Increasingly the complexity of projects creates a gamut of interested parties with something to gain or something to lose from the development of a project. These stakeholders exert such power as they possess to influence the outcome of the project. Managing the project on behalf of the client requires that stakeholders are managed effectively as a function of the project management process.

Clients and stakeholders

The definition of 'stakeholders' is less than clear. On some occasions the term is used to include any entity with an interest in a project, but such a definition encompasses those with no power to influence the project definition or its outcome. Other definitions are more prescribed, including only those with the capacity to influence the project directly. For the purpose of project management only those with the power to influence the project are of direct concern. This group can be categorised broadly as those stakeholders outside the construction process and those inside the process. In systems terms those outside are in a project's environment and those inside comprise the project's subsystems. The reason for the recognition of stakeholders more explicitly over recent years is due to the increasing turbulence and strength of environmental forces.

It is the management of the powerful stakeholders in the project's environment which primarily concerns the project management process in the definition and early stages of the project's development as they comprise organisations representing, for example, local and central government agencies, political bodies, regulatory authorities and 'green' groups. Their power can be founded on legal sanction (e.g. the planning system), influence (e.g. the political process) and disruption (e.g. the union system). If any issues of this kind exist, the project's definition can become fraught with difficulties. It is not the client that becomes complex, but the project's environment.

Entities within the construction process are often classified as stakeholders (cf. Winch 2002; Newcombe 2003), but can be seen as a different category as they do not have the potential power that those outside the process may possess. On the other hand consultants, contractors, subcontractors and suppliers are all, to some extent, likely to be subject to buyer dominance in the supply chain (Cox & Ireland 2002). Whilst they can be seen as stakeholders, they are unlikely to be a strong influence on the definition of the project. Increasingly their interests are being satisfied by new organisational forms such as partnering and other supply chain initiatives and subject to client-led initiatives for greater integration of the supply chain (Briscoe *et al.*, 2004).

Other powerful stakeholders can be conceived as being in the client's environment, but will transmit their influences from the client to the construction process (Fig. 4.4). Examples of such entities are those representing funding sources such as financial institutions and public sector funding bodies, the clients' customers and any other force acting on the client body, the action of which will influence the design and construction of the project. Client reactions to pressures to respond to sustainability issues are also responses to environmental forces, as are partnering and other supply chain arrangements, as is the client's systems response to its dissatisfaction with the traditional organisational structure of the construction industry (which in this context is part of the client's environment).

Managing a project's environment has been shown to be one of the major tasks of project management. Whilst managing the environment is much more than just managing external stakeholders, it is one of the most challenging aspects. In the midst of the fashionable focus on stakeholders, it must not be forgotten that the construction process also has a challenging and vital task in defining and implementing the client's requirements, which are paramount. Although the stakeholders' wishes may, to a greater or lesser degree, be able to be accommodated, the client is the one paying for the project and taking the risks inherent in its production and in creating value; as such, the project management process on behalf of the client places the client in a predominant position. Newcombe's (2003) belief that 'the project should be managed for the benefit of *all* its stakeholders . . . ' and that 'Project Managers bear a fiduciary (trustee) relationship to the stakeholders and to the project as an abstract entity. They must act in the interests of stakeholders as their agent, and must act in the interests of the project to ensure its survival' may seem somewhat overstated. Having clarified the position of the stakeholders vis-à-vis the construction system and its environment, this chapter focuses on the client.

Where the client is to be both owner and occupier of a building, the idea of a client appears straightforward but even in such circumstances may not be so. More frequently the client will not be both owner and occupier and the situation may be more complex. For example, the owner in the first instance may be a property developer who sells the property to an investment company which then leases it to an occupier. In the public sector the client may be a local authority which receives finance for the project from central government to implement government policy, and the users may then be a third party, for instance teachers and pupils in the case of a school. There are many more similarly complex cases and thus the first question that needs to be answered is: who is the client?

A construction project team will tend to recognise its client as the body that has the authority to approve expenditure on the project, the form that the project has to take, and its timing (and who pays the fees). The project team will find it simpler if all these authorities are vested in one body but frequently this is not so in practice. For example, central government may allocate funds for a project to a local authority, which will be responsible for developing the project, but it may reserve the right of final approval of both expenditure and aspects of design. A similar situation can occur between head office and regional office of a private client. If another group receives or occupies the building after completion, it will be the client of the client of the construction process, and the responsibility for satisfying that client will rest with the commissioner of the project. Nevertheless, in some instances, the project's client may wish to involve such a third party in approval of the design and this can further complicate identification of the client for the project team.

The members of the project team need to have the ability to understand the structure of their client's organisation, their relationship to others with an interest in the project and what makes them tick. In particular they should understand the decision-making mechanism of the client's organisation and where authority for decision lies. Only when this is known will the project team be in a position to obtain information upon which it can act with confidence.

Having understood the organisation structure of their client and how it operates, the project team will be in a position to build up the trust and confidence necessary for it to obtain accurate and useful information that will enable it effectively to develop an appropriate brief for the proposed project. Bearing in mind that every company and public authority and even every individual is potentially a client of the construction industry, the breadth of knowledge of organisations required by members of project teams is so vast as to be unrealistic. Project teams therefore need to acquire some conceptual tools such as those discussed in Chapter 3 with which to analyse and understand their clients' organisations.

However, the reality is far from such a complete understanding of client organisations on the part of project team members. It is often the case that members of the project team do not meet the client. The project team leader acts as surrogate client in many traditional contracts and the team members know the client's requirements at second and third hand only.

Clients have important responsibilities to fulfil which cannot be delegated to the project team, such as selection of project team members, setting key dates, brief development and monitoring of the project. Whilst the client may wish to delegate many aspects to the project team leader the warnings of the National Economic Development Office (1978) many years ago should still not go unheeded. It pointed out that 'the standard of service given by the building industry relates closely to the amount of effort expended by the client in establishing a good brief' and that 'satisfaction at the construction stage is closely linked to the degree of control and supervision by the client himself'. It is therefore important for clients and project team leaders to ensure that clients are appropriately integrated into the project's organisation structure.

Such sentiments continue to be echoed and found relevant (Ward *et al.*, 1991). They concluded that it is important for clients to:

- 'Set clear objectives.
- Subject objectives initially set to careful trade-off analysis.
- Consider objectives carefully when choosing a procurement method.
- Communicate objectives clearly to other involved parties and avoid conflicting guidance to different parties.
- Ensure that reaction to unexpected events involves proper revision and consideration of client objectives.'

This message was still being sent in the new millennium by Lim and Ling (2002) who, in response to a questionnaire survey, identified five attributes as contributing to project success. One was 'client sets down project objectives clearly'. The others, also equally to be expected, were 'client is creditworthy, client does not contribute to project complexity, client is not litigious and client trusts project team members'. Whilst interesting, they do veer towards utopia rather than reality. In an unusual slant on clients, Soetanto and Proverbs (2002) looked at clients' influence on the satisfaction of contractors. They found that the three key aspects which had a significant influence were 'the capability of the client's representative, the client's past performance and project management experience and the financial soundness and reputation of the client'.

More tellingly, Thompson (1991) considered that clients should be taking an even more proactive stance when concluding that:

'The owner of a project must provide clear *direction* and timely *decisions*, and must assist the project-management team to drive the project to a successful conclusion. He or she must accept the risk associated with the enterprise, and assume particular responsibility for:

- the selection of the project team,
- thorough appraisal and realism over risk,
- championing the project in the political and public arenas.'

Most importantly he pointed out that corporate client organisations are rarely suitable for providing client management of projects as the style of project management is likely to be more dynamic then that of corporate management particularly when the latter has a rigid hierarchial management structure linked to slowly changing long-term objectives. He recommends that a strong temporary corporate project team should be formed to support project management. These messages remain equally relevant today.

Classification of clients

Potential clients of the construction industry are a too large and varied group for any meaningful detailed classification to be prepared. Nevertheless an understanding of clients is aided by a broad categorisation. What follows is, initially, a simple grouping to identify the basic types of clients followed by an account of attempts to classify clients using more sophisticated approaches. Whilst the basic types of client include corporate and public sector clients separately, extreme complexity of clients is found when these two major client types are combined into one in PPP/PFI types of project.

The individual client

The individual client is the exception for most construction projects, particularly where the client is to be both owner and occupier. But even at this level the client can be more complex than expected. A particularly simple example is a couple proposing to have a house built for themselves. In such a situation there is a direct relationship between the clients and the leader of the project team, and communication of information should be straightforward. However, even at this level it can become complicated. Who is the client: the husband, who may be providing the finance, or his wife who will probably be the main user, particularly in the kitchen (or vice versa, to avoid being sexist)? Relationships between married couples vary considerably so the project team needs to understand the particular relationship, which could be difficult! This may be taking the point a bit too far, but it illustrates the problem.

A similar but amplified situation occurs in the case of the sole owner of a business. In this case the relationship between owner and employees is important. Will the owner instruct the project team alone or will the workforce also be involved? Is the owner able to indicate clearly to the project team the activities to be housed or will the workforce need to be consulted? If they need to be consulted, what does the team do if the owner is not sympathetic to their views? Even at this relatively simple level the way in which the team obtains the information it needs can depend upon understanding the client's activities, organisation and relationships.

The corporate client

The broad classification of corporate client includes all companies and firms controlled other than by a sole principal. This group therefore covers all companies from the small, simply structured organisation to the massive multinational corporation. The myriad of functions, sizes and structures of firms within this group poses particular problems for the project team. If it is to carry out its work well, the team will need to understand the objectives of the corporate client, and these will often be complex. In particular, it will need to understand the purpose of the project for which it has been commissioned and how it is intended to contribute to achieving the client's long-term objectives. To understand the objectives of the client, and to establish the firm's objectives, it will have to be familiar with how the client's organisation operates. Such knowledge is also required to identify where the best information is likely to be available on which to base the project proposal. It will also be necessary for the project team to be able to assess the ability and status of the members of the client's organisation who are transmitting the information to them.

As no two clients within this group are likely to be structured identically, the organisation analysis skill demanded of the project team is very high indeed. Coupled with this is the need for the team to be able to build up and maintain the confidence of the client, for only if this exists is the team likely to be able to obtain the information it needs to do the job effectively, much of which may be of a confidential nature.

Perhaps the only common component of such companies is that final authority will lie with the board of directors or equivalent group and in some companies it may in reality lie with only one member of the board or with a small group of directors. The leader of the project team will have to cultivate confidence at this level. Nevertheless, in the more complex company interrelationships that frequently exist, it may be that full authority does not lie with the board but with a board of another company which has control of the client's company. Such situations can make it very difficult for the project team to proceed with confidence as decisions may be overturned to the serious detriment of progress on the project. Insights into the nature of client organisations are provided by organisation theory and the work on power in organisations. For example, asking which of Mintzberg's organisation classifications fits your client organisation can be illuminating as can an understanding of the political forces at play.

Frequently the project team will have to talk to a large number of people in the client's company to tease out the brief for the project and to develop it into an acceptable final proposal. This does not have to be restricted to managers but may also include the operational people in the company. Often clients may not know clearly what they require. The briefing process requires a large number of important decisions to be made by the client and the source and authority of such decisions have to be identified. These decisions are not just about the functional attributes of the building but most importantly about the timescale and budget for the project. Of particular importance is the timing of decisions by the client on important aspects of the project, e.g. budget approvals, as the incidence of decisions will have a fundamental effect on the progress of the project.

It is not uncommon to find that the client's company will appoint a project co-ordinator or internal project team from the company to act as the link between the client and the project team. This has been found to work successfully but it does, of course, depend predominantly upon the quality of the person(s) appointed. What is of vital importance is the authority of the person(s) in this position. If members of the project team are to rely on that person's instructions, they need to be sure that they have the authority of the board to issue instructions. If they have not, then the result will probably be frustration, delay and abortive work. The ground rules need to be clearly laid down with the client's board of directors at a very early stage. This presents a very real problem as the project team is likely to be rather diffident about 'pushing too hard' for fear of offending an important client and this is often linked with a reluctance to pry into the client's affairs, although both are necessary if an acceptable project is to be realised effectively.

Dealing with a client can be a very frustrating business, particularly in the case of large clients, which tend to move towards a bureaucratic form of organisation. This can result in the project team adopting procedures that result in their designing what they think the client wants without basing it upon investigation of what the client needs. The orientation of the project team should be strongly towards finding out rather than constructing a series of cockshies for the client to criticise. This demands skills of investigation and a large measure of diplomacy.

The public client

Public clients include all the publicly owned organisations that have the authority to raise finance to commission construction work. In all such cases the funds will normally be raised by taxation or in the money market on the authority of the government. They include local authorities, government agencies and the government itself. Normally, authority to spend money on construction stems from the government but usually, when authority has been given, the 'client' may control the spending of money within certain constraints, although withdrawal of authority is not unknown.

Many of the features described above for the corporate client are applicable to public clients, particularly government agencies but the situations encountered are often more closely constrained and difficult through having to work through committees whose authority may not be clearly defined. The bureaucratic rules that surround most decisions to construct for public clients can often lead to an inefficient construction process. A common example is the establishment and approval of a budget for a project and the limitation of having to place the construction contract during a particular financial year. This approach can produce situations in which the cash flow and budget are inflexible, and so may inhibit the project team's ability to obtain value for money, particularly where virement between different expenditure headings is prohibited. Similarly, value for money may be difficult to achieve if a project has to be rushed in order to meet a financial-year deadline.

Such difficulties may have to be faced within a structure which requires that the project team is instructed by officers of the public body who are subject to control by a committee of elected or appointed representatives. They themselves may be controlled by higher level committees, either within the public body and/or within government. The process may also include the involvement of central government officials. In such circumstances the difficulty of knowing just where the decision making is done can be a severe problem.

The project team has to develop skills in understanding how such organisations work. The wide range of public clients and their objectives, many of which may be politically generated, place great demands of client analysis on the project team. Objectives can be difficult to pin down and unresolved conflicts may exist between the various client interests and the potential users of the project. If final authority in such situations rests with an elected committee, the outcome may be unpredictable and the project team will be faced with uncertainty.

As with corporate clients, there can be great benefit in working through a co-ordinator from the client's organisation provided that his authority is clearly stated and understood by both the client and the project team. The ground rules for the validity of information and decisions have to be laid down at an early stage, but even the ground rules may not be protected if there is a change in policy, often resulting from a change in composition of the elected committee who have final authority over the project.

Uncertainty and complexity stemming from the nature of the client body are compounded on overseas projects because of the project team's unfamiliarity with the client's organisation. Many of the differences arise for historical or cultural reasons. A significant amount of homework needs to be undertaken on both the influences at work in the country as a whole and on the culture and attitudes of each client.

Client profiles

An indication of the range of client types is provided by Bresnen and Haslam (1991) in their study of client attributes and project management practices. Their conclusions are that the conventional wisdom of a large number of 'naive' clients of the construction industry, who really do not understand the nature of the process and its inherent difficulties, is misplaced. They believe that the industry is one in which there are a sizeable number of regular clients whose average project is one in which they have considerable experience. Such clients typically manage a fair-sized portfolio of projects varying in size and type, and will often have some in-house capacity and well-established mechanisms and procedures for handling them.

Masterman and Gameson (1994) draw on the work of Higgins and Jessop (1965), and Nahapiet and Nahapiet (1985) in their classification of clients which they base upon:

(1) whether clients are primary or secondary constructors
(2) clients' level of construction experience.

Their definitions of constructors are:

Primary: 'Clients such as property developers, whose main business and primary income derive from constructing buildings.'

Secondary: 'Clients for whom expenditure on constructing buildings is a small percentage of their total turnover, and for whom buildings are necessary in order to undertake a specific business activity, such as manufacturing.'

Levels of construction experience are defined as:

Experienced: 'Recent and relevant experience of constructing certain types of buildings, with established access to construction expertise either in-house or externally.'

Inexperienced: 'No recent and relevant experience of constructing buildings, with no established access to construction expertise.'

When these two characteristics are considered together the following four alternative client types are produced:

- Primary experienced
- Primary inexperienced
- Secondary experienced
- Secondary inexperienced.

Whilst such a framework is useful it should not be allowed to disguise the fact that the classification identifies four basic types amongst which there exists a large variety of gradations which can be represented as in Fig. 5.1.

Chinyio *et al.* (1998) consider that the traditional classification of clients as public, private and developer clients is an insufficient classification. They propose

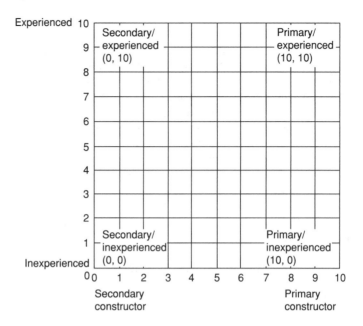

Fig. 5.1   Graduation of client types (developed from Masterman & Gameson 1994).

a classification of five groups based on client needs but cannot 'guarantee a definitive conclusion'.

An alternative approach

An important alternative to viewing construction clients according to their 'type', which he sees as oversimplistic, is put forward by Green (1996). He considers that categorising clients according to 'type', is attempting to understand clients 'from the outside' and that client organisations should be understood as social systems which requires that they be understood 'from the inside'. This can only be achieved by a naturalistic research approach.

Following Checkland (1989), he argues that 'fact and logic will never supply a complete description of a human situation. Equally important will be the myths and meanings by which human beings make sense of their world'. Hence the majority of literature relating to client organisations, which assumes that social reality is 'hard' and 'tangible', is highly questionable. The naturalistic approach requires the acceptance that there will never be any single interpretation of a complex multifaceted client and that 'It therefore becomes necessary not only to understand the characteristics of the client organisation, but also to understand the ways in which the client makes sense of his own organisation'.

Green uses these arguments as the basis for developing an approach to briefing that relies on the concept of organisational metaphors. Organisational metaphors, i.e. thinking about organisations as though they were something else, are used by practicing managers to make sense of organisational life. He identifies three metaphors for the briefing process. The machine metaphor relates to a situation in which it is taken for granted that clients' objectives are clear and predetermined and that they remain static over time. The cybernetic metaphor has similarities with the machine metaphor but with some important differences. Clients that are repetitive developers prepare standard briefs as a basis for designers. Designers are encouraged to suggest changes on the basis of their own experience to achieve continuous improvement by means of 'responsible innovation'. Organic and political metaphors relate to the need to produce briefs by means of extensive collaboration between clients and designers over time, as clients are often incapable of producing their own briefs and their needs and objectives need to be probed in depth.

It is argued that the approach chosen by professionals is often questionable due to their tendency to employ inappropriate metaphors. Designers tend to have a default metaphor that they apply in the absence of any real thought. They act in accordance with a predetermined set of assumptions and procedures that they acquired from previous experience. This is used to explain why professionals who are used to working with unitary clients often find it difficult to adjust to pluralistic clients. For pluralistic clients, the briefing process must be underpinned by more sophisticated metaphors than the machine metaphor. Green believes that in these circumstances it is necessary to construct a common understanding of the 'problem' and that design is no longer seen to be a straightforward technical process of problem solving but is seen as a social process based on iteration and learning. He also believes that since the property market became increasingly consumer-led in the 1990s, the organic metaphor has predominated and collaboration has emerged as the accepted paradigm of practice.

A subsequent complementary paper by Green (1999) makes the case for a soft systems methodology (SSM) as a way to implement these ideas in practice. The systems models of SSM are 'seen to be epistemological devices by means of which we can learn about a multi-perspective social reality' (i.e. pluralistic clients). He believes that:

'The underlying epistemology of social constructivism, together with the emphasis on negotiating a shared understanding through participation, are readily compatible with the iterative, learning paradigm of client briefing as advocated originally by Goodacre et al. (1982)'

(which is represented by the organic metaphor.)

Whilst these concepts and the language in which they are expressed may at first appear alien to many associated with the real world of project management, they are nevertheless potent tools for understanding the complexities of client organisations and the ways in which their needs have to be translated into briefs for projects.

Clients' objectives

The most important feature of any building project should be the client's objectives in embarking on the construction of the project. The need for the project will normally have arisen from some demand arising from the client organisation's primary activity. For example, the client's primary activity may be food processing. In order to remain competitive the client may need to absorb within the firm work previously subcontracted, which may require construction work to be undertaken. A client with a plant located overseas may wish to provide it with its own power supply in the event of an unstable political situation threatening existing sources. The client may be an education authority and there may be a demand for additional school places, which have to be provided by the construction of a new school. A health authority may be required to respond to advances in medical science by providing a new treatment facility.

The needs of clients will therefore be stimulated by the environment of their organisation, which will present opportunities to which they respond. Such external stimuli may be economic forces, which give the opportunity for profit, or sociological forces, which present the chance to respond to a social need, but more usually they are combinations of different classes of stimulus. The basic response of a client to environmental forces is the result of the need to survive; above this level the client responds in order to expand as a result of drive and motivation. Survival is the basic objective of clients and can be defined as maintaining their position relative to those of their competitors. This is more easily conceived for commercial organisations but it is also true for public clients. In commercial terms it requires sufficient response to stay in business. In terms of non-profit organisations, it means sufficient response to prevent the organisation being replaced by some other mechanism: for example, the establishment of urban development corporations to undertake some of the work of certain local authorities. A further example is the establishment of housing associations to provide some public housing previously provided by local authorities.

The effect of forces in the client's environment will therefore trigger the start of the construction process although it may not be realised at the time that a project is needed and at that stage it is unlikely that any member of the project team will be involved. When it becomes apparent that a construction project is needed to satisfy the client's objectives, the brief begins to be formed. A common major problem is that the project team will normally not be involved at this early stage and a number of important decisions which may inappropriately constrain the design of the project may have been made by the time they are brought in.

The client's brief is often perceived as a reasonably detailed statement of what the client requires but it is important that the strategic level of the brief is not overlooked at the expense of detail. The client will be concerned in the three major areas of price, quality and time. What is actually included in each area is for definition by each client. The quality element could, for example, include environmental and sustainability issues. The weight different clients give to each of these factors will vary. They will require value for money but on their own particular terms. A public client with a low capital budget but high revenue budget may wish to suppress the initial price at the expense of quality; a client building a prestigious building may prefer the opposite; a client in a rich, developing country may see the time for completion as paramount at the expense of price but not of quality, and so on.

Fig. 5.2   Client objectives – weighting of factors.

Quality and price can be subdivided into components, each of which will have its own weighting within the balance which a particular client wishes to achieve between competing factors. A summary of a typical weighting is illustrated in Fig. 5.2. The balance which the client requires may not be possible and compromise between conflicting factors may have to be negotiated. For instance, it would be illogical to have a low weighting for technical standards and a high weighting for low life-cycle costs. Ward *et al*. (1991) draw attention to the trade-offs which are necessary between these interdependent factors. They believe that the problem of trade-offs is in most cases complicated by uncertainty about the nature of the interdependencies between the different value-for-money criteria. They point out that the pair-wise effects for the three basic criteria of time, cost and quality do not always work in one direction and may depend on circumstances. For example a decrease in project duration may lead to an increase in cost but it does not have to, it may cause a decrease. Compounding the problems associated with establishing clients' objectives and subsequently a brief, is the apparent reluctance of architects to accept formal briefing techniques as epitomised by resistance within a student architects' design studio culture towards analytical techniques (Smith *et al*. 1998).

This type of strategic scenario will be required as a backcloth against which the detailed brief can be prepared. The client's priorities will therefore have to be established. It may well be that there is conflict within the client's organisation regarding priorities. The project team will need to be confident that it has interpreted the balance properly and to achieve this it will have to understand the client's organisation, its decision-making process and where its highest authority lies. It will be against such a concept that clients will ultimately judge their satisfaction with the completed project and upon which the reputation of the project team will rest. This is not to say that it is easy to make clients understand the conflicting pressures of a construction project. Most clients would expect each component to be weighted at 100%. Given unlimited capital this may be possible, but is not the reality of most construction projects.

In a comprehensive review of work on client objectives D.H.T. Walker (1994) summarises the lessons learned as:

- 'Clients with a detailed and firm idea on what needs to be done and how this can be achieved in a well planned and appropriately controlled manner can be viewed as sophisticated and will probably assist in a successful project outcome.
- Clients who are novices or unsure of how to brief principal consultants may be successful if they know the limits of their expertise in this matter and know how and when to ask for appropriate assistance.
- Clients must be firm in the statement and direction of their objectives and they must speak with one voice. This need has led to the rise of the role of the project manager.
- Clients should remain involved in the whole design and construction process to ensure that work is being undertaken effectively and that they can contribute positively throughout the process.'

These are the lessons learned by the investigators. The trick is how to have the lessons learned by clients and project team members. Whilst the briefing process just described is rational and makes sense as a way to proceed, a study of five projects which identified twenty examples of reasons for briefing failure (Barrett & Stanley 1999) found that the root of the failures was human nature. Good practice was 'a good thing', but in reality the recommendations were rarely followed and published briefing guides were seldom used. Experience was seen to be the driving force behind the process followed, and was seen to be personal, with no checks on whether the process was effective until gaps or errors became apparent.

The study also points to the difficulties encountered by clients who find it difficult to visualise the completed project, how it will be used by their organisation and to what extent it will improve their company. Uncertainty, therefore, is induced in the client organisation, which is often compounded by the client using the project to re-engineer their organisation. By definition, 'an organisation that requires a building is in a state of change. It will no longer be quite the same after the new building is complete'. In such a situation the internal forces of competing interests of key personnel can make the briefing process very demanding.

An approach to such problems suggested by Othman *et al.* (2004) is 'dynamic brief development'. The idea is that the brief develops over the life cycle of the project, as they found that 'the case studies client organisations continued to use change orders in order to achieve their expectations and enhance their projects' performance...'. The study recognises that changing the project brief often affects project cost, time and quality negatively, but also that changing the brief enabled clients to enhance the performance of their projects.

Barrett and Stanley (1999) make the telling point that construction professionals invariably blame the client when briefing goes wrong, with the words: 'Clients don't understand construction, they don't provide information on time and, worst of all, they change their minds!' and that 'An industry that blames its clients for the problems that exist in a key process such as briefing is doomed to failure. It will certainly not be able to improve the briefing process so long as it locates the reasons for its problems as being out of its control'.

Relationship of the client's organisation and the construction process

An organisation can be considered as an open adaptive system in terms of a general input–output model as in Fig. 5.3. Both the client's organisation and the construction project organisation can be considered in this manner.

An open system is in continual interaction with its environment and retains the capacity for work or energy transformation. A system must receive sufficient input of resources from its environment to maintain its operations and also to export the transformed resources to the environment in sufficient quantity to continue the cycle. For example, the client's organisation, whether private or public, receives inputs from society (its environment) in the form of people, materials, money, information, etc. It transforms these into outputs of products, services and rewards to the organisational members that are sufficiently large to maintain their participation. The output is therefore returned to society (the environment) in some form. The project organisation performs in the same way although the nature of the inputs and outputs differs.

In Chapter 1 it was suggested that the management system of an organisation could be seen to consist of:

- the organisation subsystem
- the behavioural subsystem
- the technical subsystem
- the decision-making subsystem.

The technical subsystem is defined by the technology required to undertake the task of the organisation and is represented by the skills, knowledge and equipment required and the way in which they have to be used. Although it

Fig. 5.3 General model of an organisation as an open system.

can be developed and adapted by the organisation, it is frequently prescribed by the current external state of development of the particular process. The technical subsystem is that to which the behavioural subsystem has to relate and with which it must be integrated. The behavioural subsystem will have a significant influence on the effectiveness of the utilisation of technology. The organisation subsystem is the structure that relates the technical to the behavioural subsystems, and the decision-making subsystem is the mechanism through which the managing system activates the organisation.

The implication of this scenario is that analysis of the technical system will produce a systematic picture of the task and of an organisation to which the other subsystems relate. Such a view of a client's organisation allows the project team to understand the *modus operandi* of its client's organisation and gives it a basis for integration during the construction process. The technical subsystem of a client can be readily perceived in manufacturing industry (e.g. manufacturing vehicles, electrical components, producing oil, etc.) but it is equally applicable to non-manufacturing activities, for example treating patients (health authority), collecting taxes (inland revenue), designing advertisements (advertising agency), etc.

This scenario also applies to the construction process itself. The technical subsystem is the technology required for designing and constructing the project. The behavioural subsystem is the attitudes and values of the members of the process. The organisation subsystem is the way in which they relate to each other and the decision-making subsystem is the mechanism through which the process moves forward. The technical subsystem dominates and in this respect it is important to recognise the differences in the technical subsystems between construction projects. For example, the technical subsystem for the construction of a house is quite different from that for a multi-storey car park, a theatre or a power station, and so on. It therefore follows that the organisation and decision-making subsystems should be designed to reflect the technical and behavioural subsystems. This demands that a variety of organisation solutions should be available to suit the particular project. There are therefore two systems involved, that of the client and that of the construction process, and they become joined temporarily for the duration of the project. The construction process becomes a temporary subsystem of the client's organisation, as shown in Fig. 5.4.

The client's primary activity can be seen as an input–transformation–output system and a response to environmental forces triggers the start point of the construction process. A part of the input to the client's primary activity (e.g. money, energy) is diverted to become an input to the construction process, which will also acquire other inputs directly from its environment. In both cases the inputs can be summarised as materials, information and energy. (Energy is the input that drives the transformation process and therefore includes people, ideas, power, etc.) The output of the construction process will then return to the transformation process of the client's system to provide an additional facility, which will contribute to the primary activity and assist in achieving the client's objectives. The construction process can therefore be conceived as an internal transformation within the client's system and as a temporary subsystem of it.

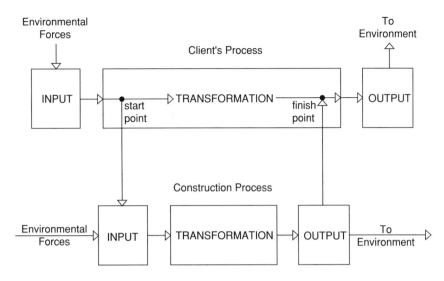

Fig. 5.4 An input–output model of the process of providing a project.

Conflicting objectives

It has been widely recognised that business organisation systems tend to have multiple objectives and that some form of compromise takes place. Multiple objectives arise as a result of the network of relationships that exist within a system. This is particularly to be expected for construction projects owing to the client–construction process relationship and the fact that the construction process itself often consists of a number of organisationally independent firms. The benefit of taking a systems view is that the conflicting multiple objective situation can be made explicit.

Multiple objectives arise because of the individual aspirations of the subsystems (e.g. firms or departments), which tend to develop their own purpose outside the main purpose of the system. It is therefore important to identify and relate a system to predominant objectives. In terms of the client–construction process relationship, the predominant objective must be that of the client, which will reflect the primary function of his organisation.

One of the tasks of the management process is to ensure that subsystems remain orientated to the primary function of the system. A company must be sure that its subsystems (e.g. departments) are not developing discrete objectives that conflict with the company's primary objective. A good example of this is what is often referred to as 'empire building', in which managers of departments are concerned only with building up their own department irrespective of optimising its contribution to the firm. The possibility of conflicting objectives in the construction process is even more likely as not only may the client's subsystems develop discrete objectives and try to implement them during the briefing of the project team, but the subsystems (e.g. firms) making up the project team may also develop discrete objectives that conflict with the client's objectives. An important role of project management will be to ensure that the objectives of the client are accurately and clearly stated and that all the contributors to the project remain orientated towards them.

The way in which objectives are established is closely associated with the power structure of the project participants which, if not controlled, can be complex and inappropriate in achieving the client's objective as discussed in Chapter 8. The project manager must be allowed to develop sufficient power to ensure that the client receives the project he requires. Although the 'information explosion' of recent years has given great focus to the power of information, recognition of the power of information has been around a long time. Many years ago Galbraith (1967) was making this point. It is essential that the client has trust in the project manager such that the manager is given the information necessary for the client's objectives to be met.

The primary function of the client's organisation will be that process which it continuously undertakes in order to survive as an organisation. The construction process has its own primary function which is providing the project, which should remain compatible with the primary function of the client. Circumstances may arise in which the primary function of the construction process temporarily becomes the primary function of the client's system but normally it will remain subservient to it. For example, if it is known that the only way by which the client's firm can survive is by completing and commissioning a building so that a particular process can come on stream by a particular date, the client's primary function will, temporarily, be to ensure that this happens. A danger following a temporary shift of the primary function is that it may lead to a permanent redefinition to the detriment of an organisation's ability to survive. Similarly, if the leaders or members of an enterprise do not agree on their definition of the primary function, the survival of the enterprise will be at risk.

Orientation of the construction process towards its objective of providing what the client requires is achieved through feedback. The client's objective for the project and the details of how this is to be achieved should be stated in the brief. Feedback loops should be designed into the construction process to establish whether the output of the process is compatible with the brief. Such feedback points should coincide with the major decision points in the process and be designed to ensure that any additional information arising from the environment during design and construction that may require the brief to be amended is taken into account.

Project change

Most construction projects are designed and constructed in accordance with the brief established early in the process, but the project team needs to be aware that uncertainty and change in the client's environment may require that alterations have to be made to the project in order to respond to them. The Royal Institute of British Architects' (RIBA) *Plan of Work* (1997), originated in 1964, indicates a time after which the project brief should not be modified. Even when completely revised in 2000 (Phillips & Lupton 2000), it stated, 'Project brief now frozen' and 'Design (final proposals) now frozen'. This is rather idealistic as it is surely not reasonable or sustainable to try and tell a client who is investing a considerable amount of money that the design cannot be changed when the operating circumstances have altered significantly. However the original RIBA *Plan of Work* recognised that in reality changes will be made but pointed out that changes of the detailed design would result in abortive work. Dynamic brief development, discussed earlier, is an approach (Othman *et al.* 2004) designed to deal with such issues.

An understanding of the state of the client's environment is necessary for the project team, since, in conditions of uncertainty of the client's needs, flexibility must be maintained. This may be achieved, for example, by not selecting a long-term fixed solution such as a new building, or by designing a building that exhibits the flexibility demanded by the environment. The implication of this is that the maximum amount of up-to-date information regarding the requirements of the client must be maintained when the project is being designed. For example, an advance in technology for a particular process (e.g. high-rise fork-lift truck design) could mean that significant amendments to the brief should be made to take advantage of such developments.

Conversely, changes in the environment of the construction process should be allowed to amend the structure of the process if this can be done to the advantage of the client. The members of the project team should keep themselves aware of any such changes and be ready to advise their client accordingly. For example, shortage of steel could result in a change of design or in ordering the steel prior to signing of the main building contract, or in bringing forward the commencement of construction by adopting innovative approaches to appointing a contractor.

Role of the client

The degree of involvement and the role of the client in the construction process will depend to a large extent upon:

- the structure of the client's organisation
- the client's knowledge and experience of the construction process
- the authority vested in the various levels of the client's organisation
- the personal characteristics of the client's people who have responsibility for the project.

If the client constructs frequently, there will probably be employees familiar with the process who can act as co-ordinators on the client's behalf and who will liaise within the organisation and between it and the project team. In such a case the client will maintain a presence close to the project. The effectiveness of this arrangement will depend on the degree of authority vested in such a co-ordinator. If this is high, it will be to the advantage of the project provided that the co-ordinator has personal skills and characteristics that gain the confidence of the project team. In these circumstances the delegation of authority to the manager of the project team by the client will not need to be high. On the other hand, if the co-ordinator does not possess much authority and/or has personal characteristics that are not suitable, this arrangement may be counterproductive as it could lead to frustration and delay.

The authority of the co-ordinator will depend upon the authority pattern within the client's organisation. If the organisation is mechanistic, it is probable that authority will not be delegated to any great degree. The project team will therefore have to rely on the higher levels of the client's organisation for decisions. This could result in delay in reaching decisions, as it is unlikely that people at such a level in the client's organisation would have sufficient time in addition to their other activities to devote to keeping close to the project. If, however, the client's organisation is structured so that authority is delegated well down the organisation, a member of the client's staff who is intimately involved with the particular project may have authority over most matters. This should result in close integration of the client organisation and the construction process, with the effect that decisions can be readily obtained to the advantage of that process.

Much has been written about the amount of authority which the 'project manager' should or should not be given. In the context of a project manager who is external to the client's organisation, this is a difficult decision for the client. The project team will be spending a large amount of the client's money and human nature does not normally allow some external agency the authority to do this. Nevertheless, in such complex circumstances as a construction project, such decisions have to be faced and resolved. Much will depend upon the structure of the client's organisation and its experience of construction. As referred to earlier, if the client appoints its representative from within its own organisation and gives that person a large degree of authority, then the manager of the project team should not require much authority as there will be easy access to the client's representative. On the other hand, if the client keeps authority for the project high within the organisation, vesting significant authority in the manager of the project team should be considered. The problems start to arise when neither of these situations exists.

Similarly, problems will arise when authority is not clearly defined as often happens when the client fudges both authority and responsibility for the project within its organisation. This usually 'knocks on' to an unclear statement of the authority of the manager of the project team. Problems then manifest themselves as either decisions not taken at the right time or even not at all, or in decisions being frequently changed leading to delay and abortive work.

Winch *et al.* (1998) showed that a proactive internal client project management function greatly contributed to the success of the Glaxco project. Also, Bresnen and Haslam (1991) found that project performance was better when the client had an in-house architectural team.

There is no set solution to the integration of the client and the construction process. Each mechanism will have to be designed to suit the particular organisation of the client body. The client cannot be expected to change the organisation structure fundamentally because of a temporary embarking on construction. Therefore the mechanism should reflect the organisation, yet clearly state the pattern of authority and responsibility for the project. Only from this basis can the decision-making process and communication system for the project be identified. The integration of the client and the project team is a most significant factor in the success of the project and requires understanding and skill in its design.

Kometa *et al.* (1994) show that certain attributes of clients other than client integration contribute to a successful project performance through the manner in which clients' attributes influence the performance of consultants. The attributes identified include financial stability, feasibility, feasibility of the project, past performance of the client, project characteristics and client duties (which did not include integration). This is of course a different perspective, concerned with the substance of the client body rather than management, but is nevertheless interesting. Subsequently, Kometa *et al.* (1994) found that clients themselves, and also consultants, perceived planning/design, project finance, project implementation/ management, and project definition/formulation as the four most important client responsibilities but with different rank orders. The study also confirmed that the degree of responsibility clients accept is a function of their experience with the industry. However, neither study examined the state of integration necessary for the clients' needs to be met. Complementing these findings were those of D.H.T. Walker (1994) which discovered that project team confidence in the client and the client's representative rather than vice versa reflects the difficult role a client's representative has to play as a link-pin between a multidimensional client group with conflicting goals and the project team.

Clients and sustainability

This section is included to show the influence of clients in transmitting (or not) the effects of environmental forces to the construction process. As discussed earlier, sustainability is a recent major issue that has emerged as a powerful environmental force on the construction industry. Although the industry needs to respond to the sustainability forces acting directly upon it, the main impact is not directly on the construction industry (Myers 2005); rather, it is transmitted from the client in the manner discussed later and shown in Fig. 4.4. The extent to which it is transmitted at the present time is uneven due to the extent to which the client organisation is required to respond. This situation is illustrated by the use of design tools to help professionals design greener buildings. Shiers *et al.* (2006) believe that 'Specifiers would only use such tools either when instructed to do so by clients or because of the specifier's own organisational green agenda'.

The general situation is well illustrated by a paper which focuses specifically on delivering sustainability through value management (Abidin & Pasquire 2005). Practical barriers to responding to sustainability issues are seen as a lack of time to consider the vast issues of sustainability and the absence of formal guidelines for sustainability. Behavioural barriers include the lack of knowledge and awareness of sustainability and passive and negative perceptions about integrating it into the value management process. Whilst relating to value management, these issues act as a proxy for attitudes generally. In particular, value management practitioners argue that value management only responds to clients' demands, but that 'although sustainability would improve the value and quality of the output, clients were perceived as less interested in sustainability during value management' (Abidin & Pasquire 2005).

What then is needed for clients and the construction industry to respond to this environmental force? If there is no response, transaction cost economic arguments would be that the market has determined that it is not economic to respond. What then needs to change to make clients respond? In the public sector, social pressure to which government needs to respond will be a key factor. In the private sector, Lutzkendorf and Lorenz (2005) argue that the incorporation of environmental and social issues into property valuation theory and practice will contribute to synergies between sustainable design and risk management, with implications for property risk assessment for lending and insurance purposes.

Increasing awareness of such issues will eventually lead to clients realising that there is a pressing economic as well as altruistic reason for responding to the sustainability imperative. Whilst the discussion above focused on commercial reasons, legislation, regulation and public opinion will also act on clients' need to respond. Governments and the European Union have enacted legislation which private as well as public sector clients have to comply with and which is bound to increase in scope and impact. As long as sustainability continues to be viewed ambivalently by clients, and until the commercial world recognises its worth and legislation is rigorously imposed, the construction industry itself is unlikely to respond with real purpose.

Nevertheless, case studies of six construction organisations (Fergusson & Langford 2006) found that:

> 'As environmental strategies are developed, competencies in managing environmental issues will grow and lead to improved business performance. However, the growth in the organisation's environmental competence provides the opportunity for increased competitive advantage.'

The advent of sustainability has presented a unique opportunity to observe a new and powerful environmental force come into being and the slowness of the response of both the commercial and non-commercial sector of society to such a pressing issue. The fact that it has enormous implications for the construction process has illustrated how systems resist environments' forces until they are forced to respond in order to survive.

Clients and projects

Project teams tend to start developing projects assuming that the client has:

(a) identified the best means of achieving its objectives
(b) carefully analysed the spatial, technical and performance requirements associated with its objectives.

The data provided by the client is therefore frequently accepted without question as the basis for developing the design. The inappropriateness of such assumptions is illustrated by the redevelopment of a ferry pier in Hong Kong. Redevelopment was considered appropriate as the surrounding area had been redeveloped including reclamation of land which was subsequently developed with residential units. As a result a new transport complex was built, into which the ferry pier was fully integrated. It was decided that some commercial facilities should be made available to attract passengers. The resultant design, which was quite novel and the first of its kind in Hong Kong, comprised the ferry pier portion that could accommodate simultaneously two triple-deck ferries and two hovercraft. The pier incorporated two further levels, one for a shopping arcade and public promenade and the other for a restaurant. The 'assumption', without detailed study, was that the demand for ferry service would increase in view of the proposed residential developments on the new reclamation. Unfortunately, because of a new subway line, the actual demand for ferry service since the ferry pier was commissioned was far from that assumed. As a result there has been insufficient patronage of the commercial development and arguments resulted over lease conditions. The commercial areas proved not viable and were eventually converted to offices. The world is littered with similar examples of lack of objectivity and no matter how efficiently resources are applied in devising and executing designs and construction, if they are not achieving realistic objectives the inevitable result is waste.

Whilst lack of an objective evaluation will invariably lead to unrealistic objectives, the internal politics of the client organisation can contribute equally to a lack of objectivity, distortion of objectives and potential problems for the project team. This process was well illustrated by Cherns and Bryant (1984):

'Consider the case of the corporate client that always has more projects competing for finance than it has capital available. Thus the agreement to invest in X (say, building a new warehouse) is a decision which not only pre-empts X^1 (extending an existing one) but also gives priority over Y (building a new laboratory) or Z (installing a computerised production system). Since Y, Z and X^1 are projects in competition with X, and all have their organisational supporters, the decision for X is a victory for the supporters of X – and a defeat for the supporters of the other competing projects. (Examples of client interest groups supporting competing projects include operating divisions, specialist departments, political factions, professional groups, etc.) The victory for X then marks a shift in the balance of power within the client organisation (if only to confirm the dominance of one interest group over another). But the victor has now given a hostage to fortune. In fighting the battles, he has probably shaded the risks and been optimistic in his estimate of cost in order to get his budget just below the organisation's threshold of approval for capital expenditure. He has promised a return and now has to deliver – or lose credibility.

The study has provided us with vivid glimpses of the various "hostages to fortune" that the different interest groups within the client complex may have offered in promoting a particular project in competition with others. Each participant can be seen as bringing to the table his own sense of what is at risk personally, as well as what is at stake professionally or departmentally, in the forthcoming project experience. Some client participants have a high stake in meeting the target completion date; others in working within the promised budgetary limit; others in the operating performance of cherished design concepts. Many of the stakes are reputational (e.g. the operational manager's reputation as a skilled negotiator in the best interests of his own department; the project manager's reputation as a hard-headed realist who gets results in the face of whatever obstacles).

In considering the role of the client in construction, then, we cannot treat the client as unitary; nor we can ignore the events which preceded the decision to build. The progress of a construction project involves various groups within the client organisation whose interests differ and may be in conflict, and whose observed behaviour cannot adequately be explained without reference to the past.'

Newcombe (2003) draws attention to the idea of the 'shifting multi-goal coalition' in which 'the power base of the main actors themselves "shift" over time', creating complex organisations in which conflicting objectives are rife. He believes this is particularly prevalent in project organisations, especially when there are many stakeholders involved which generates a 'very strong political flavour to the project process'. Conflicting objectives and political forces within the client organisation are likely to be the most powerful and of primary concern to the project team.

Such tensions are a recurring theme and are likely to be ever-present in client organisations, as often referred to in the business press and well illustrated by Staw and Ross (2005), and Royer (2005) with chapters on 'Knowing when to pull the plug' and 'Why bad projects are so hard to kill'.

It is important that project teams not only find the client's objectives realistic, but also that they have some understanding of the organisational dynamics which brought them forward. In order to be reassured on the former and have knowledge of the latter there needs to be a high level of trust and compatibility between the client and project team leader. The team leader will need to ask the client many searching questions before the brief is fully developed. It is debatable whether many clients will either be prepared to or be in a position to satisfactorily answer such questions. As a result many client objectives are unsatisfactory and lead to unsatisfactory projects for which the project teams are likely to carry a large part of the blame if not the responsibility. There is a great need for a high level of understanding of this process on the part of both clients and project teams.

Cherns and Bryant's study identified a number of tentative propositions about client involvement in the construction process of which the following are particularly pertinent to the points developed above:

- Most client systems are very much more complex organisationally (in terms of who wants the building, who will use it, who approves it, who controls the money, etc.) than is commonly acknowledged by project teams.
- Members of project teams seem to be impatient of this complexity, sometimes even embarrassed by it, and insist on dealing with a single client

representative within whom all the internal politics of the client system can be contained.

- Many of the problems concerning design changes, delays, and difficulties during the construction phase have their origins in the unresolved conflicts within the client organisation and are exacerbated by too early an insistence on an oversimplified client representative function.
- The earliest decisions taken by the client system have more influence over the way the project organisation is formed and its subsequent performance than those taken later.
- The reasons for these earliest decisions have their origins in the client's organisational culture, procedures and structures; they are often idiosyncratic, shaped by social and political forces as well as by economics and technical considerations, and may be unduly constrained by residues of the client's pre-project history.
- Any serious attempt to understand and hence to disseminate usable knowledge about the client role in the organisation and management of construction projects must take into account not only technical and economic factors but also the social and political forces acting within the client organisation, especially the influence of the client's pre-project history on the decision to build.
- The decision to build is a large-scale innovative decision with consequences for existing patterns of resource sharing and risk taking in terms of power conflicts and political behaviour within the client organisation.
- These conflicts and behaviours can critically affect the formation, development and subsequent performance of the project organisation which is set up to manage the project and of which the client system is an initiating component.

These propositions clearly illustrate the issues facing project teams. Whilst in most cases the problems they generate are unlikely to be resolved, their importance lies in project teams knowing of their existence and being prepared to understand and adapt to the dynamics at work to the benefit of the project outcome.

One particular project analysed by Walker and Hughes (1987b) is illustrative of the organisational complexity of client organisations. The project was a divisional police headquarters constructed on a confined town centre site. Technically, the client was a Metropolitan County Council. In reality the client was multi-headed and, in addition to the Council, consisted of the Home Office, the Police Committee, the Chief Constable and representatives of the police force. Compounding the difficulties of dealing with a complex client body was lack of sound integration between the client body and the project team. The complexity of the client body is illustrated by the fact that 40 *different* management units, many of which arose from the various branches of the client organisation, had, at some stage of the project, the responsibility to approve some aspect of the project. There were also 30 different units with the power to make recommendations. The completed project was unsatisfactory due in large part to the complex nature of the client which generated procedures which imposed a 'dead hand' on the project.

Allen (1984) illustrates the benefit of the project team adopting a more positive and creative approach to defining the client's needs. He does not commence from the more conventional standpoint of both client and project team which he believes is:

- 'The client has already decided that he needs a building and has already identified his site.
- The client draws up a schedule of his requirements for that building, both in spatial and performance terms. Sometimes, a layout of his spatial requirements is also provided.
- The design team produces (maybe alternative) outline proposals, in response to the client's requirements, complete with an indication of specification, programme and cost. This may, or may not, meet with the client's approval. If it does, the project proceeds through its remaining stages. If it doesn't meet with approval, it may have to be modified, to varying degrees, or started all over again.'

This process assumes that the client knows precisely what is required and all the project team needs to do is work out the details and arrange for the project to be constructed. Allen believes that it is normally wrong to make such an assumption. He illustrates the benefits of a more dynamic approach to clients' needs using the following projects:

(i) 'The client approached his project team with the intention of extending an existing laboratory complex in two places; continuing its tendency to 'grow like Topsy'. Upon investigation, it became apparent that the client had given little consideration to his future needs; both in terms of expansion and his day-to-day operations. The result of the project team's initiative was an overall development plan, involving a rolling series of building phases. Some of these included refurbishment of the existing buildings, to accommodate changing occupational patterns as the development plan unfolded. In operational terms, the entire laboratory complex became much more effective and efficient. Inevitably, this led to relatively greater profits for the client – which, after all, was his original objective.

   Had the design team adopted the more traditional approach to this particular client's initial brief, the two extensions proposed would have only served to further compound the relative ineffectiveness and inefficiency of the existing complex.

(ii) A major vehicle manufacturer approached his project team with the intention of building a new engineering centre for the whole of his car design activities. The client's initial intentions were to build a separate office block, workshop block, styling centre, etc., to house his existing departments. Adopting the approach advocated, the design team sought answers to the question 'How are (or should) motor cars (be) designed?' The building design, which emerged from this process, was substantially a single building, with each of the functions involved in designing a motor car sensibly related. The other consequence of the process was that the entire organisational structure of the company's engineering centre would change, to reflect the clarity which the briefing process had produced.

Again, this is an example of the design team seizing the initiative, in the interest of satisfying the client's real objective. In this instance, the effect of that initiative extended far beyond their immediate function as a design team.

(iii) By way of contrast, a higher education body, which, in spite of stringent cash limits, sought and achieved a similar level of service to the public as was achieved elsewhere, where cash was not so limited. The design team's approach was to give careful consideration to the use of space. By careful timetabling and the resultant greater use of each space, the total amount of new space required was significantly reduced. This, inevitably, reduced the capital investment required, without loss of service to the community.

This serves as an example of where the client had a very clear objective and had the wisdom to involve the project team in identifying the most economic means by which that objective could be achieved.'

The key to success, in each example, was the initiative and perception of the particular project team or, in the latter case, the client. In all three examples, had a more traditional approach been adopted, it is extremely unlikely that the client's needs would have been met by the resultant designs.

Peter Morris (1994) describes many of the world's 'mega-projects' in his historical account of project management including Atlas, Polaris, F-111, Apollo, Concorde, the Trans Alaska Pipeline and others. The role of the client (or 'owner' which he prefers) is strikingly portrayed in his accounts in which project definition plays a crucial role for either good or ill, particularly in cases when the client is that most complex of 'owners' – government.

These examples illustrate the complexity of the relationships between clients and project teams in which many millions of pounds are frequently involved. A successful project inevitably means that both the client and the project team leader have to work in a spirit of trust, openness, collaboration and creativity to identify the appropriate objectives for the project and so give it the greatest chance of success.

Chapter 6
The Project Team

The firms

Reference was made in Chapter 1 to the large number of firms involved in the construction process and in Chapter 3 to the theoretical basis of their formation. These firms are often independent units which are interdependent in terms of the work they undertake – the design and production of construction work. Even so they vary considerably in the range and quality of skills they offer. The number of different combinations of firms that may be involved in a construction project is extremely large. As a result, firms have to be familiar with working with a variety of other firms, and at any one time within a particular firm many different amalgams of firms will be working on the projects in hand.

Table 6.1 gives some examples of the types of amalgam that may exist, and the additional variety introduced by the various ways by which the construction contract may be awarded. The table illustrates eighteen primary examples of different arrangements, but many more variations are available. Each of the different arrangements generates a different set of relationships between the contributors. At any given time a person can find themselves involved in a variety of situations as it is not unusual for someone to be working on a number of projects simultaneously. The complexity of the situation is compounded by the variety of clients and projects which overlie the professional relationships.

Within the amalgams shown in Table 6.1, management takes place at various levels. In each case the individual contributing firms have to be managed. The partners or directors will be concerned to maximise the efficiency of their firm while at the same time enhancing its reputation for service. The same will also apply to departments of public authorities. If firms are concerned with more than one skill, e.g. a professional consortium, this will involve managing not only the individual skills but also the collective skills of the members. Within individual firms, the service provided to a particular project will have to be managed within the context of the total firm. Resources will have to be allocated to satisfy the demands of the project and also be within the capacity of the firm, and decisions regarding both the quality and quantity of resources allocated must be made. For example, the services engineer will have to manage the services engineering provisions within the resources of the services engineering practice, the contractor the construction process, etc. At a lower level each individual contributor will have to manage its own contribution. All these activities will

Table 6.1 Examples of amalgams of firms.

Designers / Contractor	Appointed by competition after design is substantially completed	Appointed by negotiation or competition early in design process	Management contract	'In-house' to client	Design-and-build
General practice surveyor, architect, quantity surveyor, structural engineer, service engineer, all in separate private practices	X	X	X		
As above, plus a project manager in separate practice	X	X	X		
Consortium of design skills including project manager and separate general practice surveyor firm	X	X	X		
All design skills 'in-house' to client (e.g. local authority, private developer), including project manager	X	X	X	X	
Some design skills, including project manager, 'in-house' to client with others in separate private practices	X	X	X	X	
'In-house' to contractor					X*

*In addition the client may appoint consultants to oversee the contract on its behalf have an objective in terms of satisfying the client, but they will also have other objectives. The dominant one will be to ensure a profit for their firm (this also applies to public authorities in terms of effectiveness) while at the same time maintaining the firm's reputation. Individuals will have personal objectives (e.g. self-advancement, avoidance of responsibility). While the objectives of the firm, the project and the individuals continue to be satisfied simultaneously, all will be well. However, if these objectives clash, it is the role of the manager of the project on behalf of the client to resolve the conflict in the client's interest.

Project management is the management of the contributors to the project who will be from different firms except where they are totally 'in-house' to the client's organisation. Its sole objective is the satisfactory completion of the project on behalf of the client. It therefore normally crosses firms' boundaries and for its purposes temporary management structures are created for the duration of the project. They will be disbanded on completion of the project but may be reformed for future projects. It is important that the contributing firms recognise the existence of temporary management structures and organise themselves so that their members become full members of those structures. This requires that firms be prepared to dedicate staff to projects even though this may at times appear to be to the detriment of the internal organisation of the firm. Firms should therefore be structured to allow staff to give allegiance to the projects on which they are employed and to be involved with only the number of projects that allows them to devote sufficient time to each. Dedication to projects should enhance a firm's reputation, and dedication to firms should enhance profit in the short run. If both are not achievable, a trade-off between them takes place to the detriment of one or both. Such a situation demands sensitive and skilful management if both reputation and profit are to be optimised and if staff are to be motivated and retain job satisfaction. Staff should be seen to be rewarded for achieving the appropriate balance.

If this balance is not achieved, the project team has been described (Association of Project Managers 1984) as having 'a limited objective and lifespan, and therefore with a built-in death wish'. The same publication also saw it as potentially a 'weak system compared with the continuous and self-perpetuating drives of the other systems. Putting together effective teams in such circumstances from a wide variety of organisations and motives is therefore a difficult and complex task'. The general management literature refers to the structure which emerges as a matrix and its management as matrix management. Mainstream management writers tend to find a matrix, and the resultant need to manage teams, difficult to relate to, with comments such as 'much ambiguity must be tolerated and competing claims accommodated for the matrix to function. For many participants, matrix structures are high demand, high stress work environments' (Scott 1992). Even now, at the beginning of the twenty-first century, teams are seen as a new innovation by the general management literature. They are described as 'one of the major forces behind today's revolutionary changes in organisations' (Schermerhorn *et al.* 2004) and project teams were still being seen as 'another type of team that has gained popularity in recent years' (Shriberg *et al.* 2005). More common classifications of types of team in general management include problem-solving teams, self-managed teams, cross-functional teams and virtual teams (Robbins 2005). These are invariably intra-organisational teams, as opposed to the mainly inter-organisational teams found in construction, although it would appear that the general management literature is recognising the distinctive nature of project teams for intra-organisational tasks. For example, Scott (2003) states: 'In some organisations work is organised so that project teams become the basis for the organisation of work, replacing functional departments and even prescribed job descriptions' and 'In some arenas, project groups become more salient than host organisations', causing Barley and Kunda (2001) to raise the question of 'whether a theory of post-industrial organising might usefully reconceptualise firms as contexts for projects'. Yet, construction professionals, contractors, subcontractors and clients have always existed in such structures as they represent the basic structure which is inevitable as a result of the task to be performed.

Criticisms of the matrix structures are cited in Poirot's (1991) paper as over-complexity, difficulties of quality control and operational difficulties particularly when applied in large companies. But herein lies the difference as much of the mainstream management literature focuses on matrix management within a single organisation rather than that required by construction project management which relates to inter-organisational management. Nevertheless Poirot continues by showing that in his own multidisciplinary consulting firm the matrix structure has been most effective and that:

> 'Almost any organisation structure will work if the users accept it, but with the matrix system, managers must have clear definitions of authority and responsibility. They must have mutual respect for each other and must be excellent communicators. They must be able to set aside their individualism and understand that the essence of a matrix is that the group must succeed first, then the individual.
>
> The matrix system is a power-sharing, power-balancing organisation. If all the power moves to one side of the matrix, the whole organisation loses. Most decisions are made at lower levels of management in order to push decision making as far down as possible in the organisation and to encourage participative decisions.
>
> Our matrix works! We feel the decisions creating our matrix have resulted in continuous growth and the ability to weather recessions without major decreases in our permanent staff.'

Further insights to the differences between inter-organisational and intraorganisational matrix management are provided by Robins (1993). He drew attention to the fact that in-house project managers always have the authority to control subcontract expenditure (which can also be interpreted as consultants in construction) when work packages are subcontracted out but not when the work is undertaken by in-house functional departments. In the latter case he believes that separation of responsibility and authority occurs and that effective management control is prevented. He believes that if a functional department is seen as a subcontracting business within the organisation and accounting procedures are set up to recognise this, then the problem will be solved.

It becomes the job of the in-house functional managers (subcontractors) to sell their services to the project manager against competition from external subcontractors (or consultants). These views highlight the difficulties of in-house matrix management as opposed to matrix management of external organisations. Although the latter overcome many of the criticisms in the general management literature they can of course bring problems of their own such as contract disputes and adversarial attitudes.

The general management literature refers to both 'teams' and 'groups', but does not always make clear the distinction between them. Robbins (2005) provides useful definitions, stating that a group 'interacts primarily to share information and to make decisions to help one another perform within each members' area of responsibility' and a team 'generates positive synergy through coordinated effort. Their individual efforts result in a level of performance that is greater than the sum of those individual inputs', the latter clearly reflecting systems thinking. The confusion between groups and teams is well demonstrated by Kakabadse *et al.* (2004), who state that: 'A team must think of itself as a group'.

Much of the general management literature on groups and teams may not seem to be directly relevant to construction organisation, as it deals with aspects which can be influenced by intra-firm management but which are difficult to achieve in inter-firm organisations, even by the better project managers. For example, group structure requires people to be included for their talents as implementer, innovator, monitor, shaper, etc., and not for their specialist skills as in construction project teams. Other aspects are equally difficult to transfer, such as group norms (expectations), cohesiveness (group identity) and reflexivity (assumption of certain 'givens'). Such aspects of groups and teams that are in-house to an organisation can be identified and managed by the organisation, but the task of managing construction project teams whose members are drawn from diverse firms, each with its own expectations, culture and way of working, is a much more challenging business.

However, whilst the above ideas from mainstream management are only of general interest to construction project management, team building is one aspect of the literature on teams arising from general management which could with benefit be developed by construction project management. Team working does not occur naturally, as many people do not take naturally to being part of a team; many prefer to work independently. This can be particularly so in the specialist areas of construction. As team building is participatory, the most important aspect is for the team as a whole to recognise when it is not working effectively and then participate actively in deciding what needs to be done to improve the situation. There are a number of mechanisms in use to achieve these ends, such as formal retreats and 'outdoor adventures' of various kinds, many of which are received with scepticism and cynicism, particularly by those in the construction industry. Whilst these activities may not be appropriate to construction, team building is rarely treated seriously by construction project managers and ways need to be devised to enable project teams to learn how to work more effectively as integrated teams.

Relationship with the client

The complexity of project management structures raises the issue of how best the project team should be integrated with the client's organisation. The two ends of the spectrum are illustrated by, at one end, the project manager or other leader of the project team as the only point of contact between the team and the client, with all instructions and advice being passed through this channel. At the other end, all members of the project team have direct access to the client and in this arrangement the leader of the project team co-ordinates the instructions and advice given. Either of these alternatives is likely to be unsatisfactory in most cases and the appropriate integrating mechanism will probably lie somewhere between.

The design of the mechanism will depend to a large degree on the amount of authority delegated by the client to the manager of the project team. Where substantial authority is delegated, most of the contact between the client and the project team is likely to be directly with the manager of the project team acting as the surrogate client. In such cases, for aspects for which the manager of the team does not have authority to act, the client may well prefer the leader of the project team alone to present the recommendations. In cases where the manager has little authority, the client may prefer to thrash out issues in consultation with all or some of the members of the project team. Alternatively, the client may require the project team manager to make recommendations for a decision, so that the client can discuss them with the whole group. In this arrangement, although the manager may not have delegated authority from the client, significant authority is gained from being in the position of co-ordinating and influencing the proposition(s) that form the basis of recommendations to the client. In putting propositions and recommendations to the client, the project team manager will also be in a good position, even in the presence of other members of the team, to influence the final outcome.

Where there is direct access by team members to the client, the team manager will need to ensure that the client receives a balanced view and that decisions are made in the light of all the factors affecting the project, rather than as a result of the statements of the strongest personality in the team. For example, many construction projects do not have a formal, detailed brief. The detailed brief emerges through the architect placing a series of sketch designs before the client, which are amended or rejected. The 'brief' therefore proceeds incrementally until the client 'sees what is required' on the drawings. In such a process it is essential that the other members of the team be present and involved. Otherwise important elements, such as cost and time constraints and certain significant elements of design (e.g. aspects of services), may be ignored through concentrating attention upon other aspects. The result could be that a building is proposed which the client cannot afford and which, by that stage of design, cannot be amended in time to be completed to programme. Such a development is potentially more likely if the architect or other professional contributor is also the project team manager than if the leader were independent of the professional contributors.

There can be no hard and fast rules laid down for integration with the client. So much will depend upon the particular views held by the client and the client's experience of construction projects. However, the project team manager can influence most clients and should ensure that whatever is devised is clearly laid down and understood by everyone involved, particularly the client. The essence of integration is that the decisions made as a result of contact with the client are controlled in terms of the objectives of the project. Unilateral decisions made by either the client or one of the team can, at best, lead to confusion which will take a great deal of unravelling and cause abortive work. At worst they may be incorporated into the project, with the result that whereas they may satisfy one aspect of the project's objectives, they defeat one or more of the client's other objectives, which in the long run may be more significant to the client's satisfaction with the total project.

Differentiation, interdependency and integration

The specialisation of the contributors to construction projects has been occurring throughout the world since the last century. As observed in Chapter 2, in the UK from the basis of architect/builder have evolved quantity surveyors, various specialist engineers, general contractors, specialist subcontractors and general practice surveyors. Even within these specialist occupations there are often further specialist subdivisions. For instance, there are design architects, detailing architects and job architects; in the quantity surveying field there are building economists, those who prepare contract documents and final account specialists. Whereas on some projects the same person may undertake all the functions of a particular contributor, for many projects a number of people are involved at the different stages of a particular specialist contribution. Add to this the way in which clients' organisations are often subdivided into specialist groups, all of which have a contribution to make in terms of project definition, and the complexity of the interrelationships that emerge is clearly evident. In systems terminology such specialisation is called differentiation (Lawrence & Lorsch 1967), which can be defined in construction terms as the difference in cognitive and emotional orientation among contributors to projects who offer specialist skills. The differences in cognitive and emotional orientation of the specialists within the construction process are readily apparent. Each of the specialists tends to view his or her colleagues with a certain amount of scorn. The contractor rarely expresses respect for the architect and vice versa, and no one has a good word for the quantity surveyor or consulting engineer! The divisions are wide but narrowing; nevertheless, at project level there exists a pressing need to ensure that such differences are reconciled so that they do not affect the performance of the project team to the detriment of the project and hence the client.

Closely related to the concept of differentiation is the concept of sentience, developed by the Tavistock group (Miller & Rice 1967). A sentient group is one to which individuals are prepared to commit themselves and on which they depend for emotional support. In the construction process this can arise from allegiance to a firm or to a profession or to both. It is a particularly strong force in the construction process and it is from sentience that the perception of the process by the different contributors arises. It has been found that sentience is likely to be strongest where the boundaries of a sentient group and of a task coincide. This is the usual situation in the construction process. For instance, traditionally, architects have normally been solely involved in architecture and builders in building, with very little, if any, overlap. The various contributors have a tendency to focus upon and be concerned only with their own specialism and are unable to perceive and respond to the problems of others.

Sentience is weakest in a group of unskilled or semi-skilled workers whose roles are interchangeable and where each individual is dispensable. Individuals in such a group will not acquire sentience unless the group finds supplementary activities through which members can make individual and complementary contributions. It has been found that sentience is strongest in members of a professional body that confers upon its members the right to engage in professional relations with clients in which task and sentient boundaries coincide. There is a specific danger when both direct relations with clients and coincidence of boundaries of sentient and task groups occur in that it may produce a group that becomes committed to a particular way of doing things. Although both efficiency and satisfaction may be greater in the short run, in the long run such a group is likely to inhibit change and behave as though its objective had become the defence of an obsolescent method of working. This view appears to have some significance for the construction process. There have been many pleas over the years for the boundaries between the professions of the building industry to be broken down but there is still little evidence of this having taken place on any appreciable scale. An aspect of sentience which is rarely recognised is that to a project. Dainty *et al.* (2005) report the affinity of a project team to a cancer research facility. It was found that attachment to the project's goals positively influenced the commitment of those involved.

A phenomenon which has similarities with sentience is the concept of groupthink (Janis 1983) which is a process which can work against effective group decision making. Whilst conformity in groups (which encompasses project teams) which are charged with decision making is necessary to enable progress to be made, excessive conformity interferes with constructive critical analysis and leads to ineffective decision making. Weak arguments and uncritical thinking go unchallenged in the group in order not to disturb the stability of the group. The members of groups which turn out this way subject themselves to social pressures, self-censorship, illusions of invulnerability and unanimity, rationalisation of decisions and have self-appointed mind guards. Project managers need to guard against groupthink in their project teams. Methods of counteracting groupthink are available (Leigh 1983), one of which is 'teamthink'. Teamthink is directed at self-managing teams and compares with groupthink as follows (Manz & Neck 1995):

Groupthink

Description: Group members striving to agree with one another, overwhelms adequate discussion of alternative courses of action. Defective decision-making results.

Symptoms:

- Direct or social pressure against divergent views
- Self-censorship of concerns
- Illusion of invulnerability to failure
- Illusion of unanimity
- Self-appointed mind guards that screen out external information
- Collective efforts to rationalise
- Stereotyped views of enemy leaders
- Illusion of morality.

Teamthink

Description: Groups engage in effective synergistic thinking through the effective management of its internal dialogue, mental imagery, and beliefs and assumptions. Enhanced decision-making and team performance result.

Symptoms:

- Encouragement of divergent views
- Open expression of concerns/ideas
- Awareness of limitations/threats
- Recognition of members' uniqueness
- Recognition of views outside the group
- Discussion of collective doubts
- Utilisation of non-stereotypical views
- Recognition of ethical and moral consequences of decisions.

The ideas underpinning sentience and groupthink can be seen as part of the broader field of organisational culture which emerged in the 1980s and which is considered later in this chapter. Whilst studies in organisational culture are of interest to construction project management, the more specific concept of sentience and the practice of groupthink can be seen to be more directly relevant to management of construction projects.

The differentiation of skills together with their reinforcing sentience can be clearly seen in the construction industry. It is also quite clear that all the contributors, each in their own 'box', are interdependent in carrying out their work of producing the completed project to the client's satisfaction. The network of interdependencies is practically total. It is not that each contributor is dependent on one other but that all contributors are in some way dependent upon all the others. If members of the process were asked if they were interdependent, they would undoubtedly agree, but this is not something that would be at the forefront of their minds if the question were not asked. This lack of recognition of interdependency begins with the education of members of the construction process. Each discipline is educated in relative isolation from the others. Exceptionally there may be some joint work, but if so it is only likely to be a very small proportion of the time devoted to study. The difficulties of resolving this situation are compounded by different patterns of courses and a lack of will to break the mould and reform the educational process. This problem is particularly apparent in the UK and was referred to once again in the latest government report (Latham 1994), but is common throughout the world.

The contributors are interdependent because on the one hand the various tasks that have to be undertaken to achieve the finished project require an input from a range of contributors, and on the other hand because the tasks themselves are interdependent as frequently a task cannot be commenced until another has been completed or unless another task is undertaken in parallel. Different types of interdependency exist and have been classified as pooled, sequential and reciprocal (Thompson 1967).

Pooled interdependency is basic to any organisation. Each part renders a discrete contribution to the whole. The parts do not have to be operationally dependent upon or even interact with other parts, but the failure of any one part can threaten the whole and therefore the other parts, for example the decentralised divisions of a large, diversified company. In the construction process, if one part fails, it will not necessarily mean the failure of the other parts but the failure may reflect upon the reputation of the other parts. *Sequential interdependency* takes a serial form. Direct interdependency between the parts can be identified and the order of the interdependency can be specified. For example, bills of quantities must be prepared before tenders can be invited (using this particular form of procedure). *Reciprocal interdependency* is when the outputs of each part become the inputs for the others and the process moves forwards through a series of steps. Each step requires interaction between the parts and each part is penetrated by the others. This is seen, for example, when preparing an outline proposal for a building which must be functionally and aesthetically sound and also feasible from a structural and cost point of view.

The three types of interdependency can be arranged in ascending order of complexity as pooled, sequential and reciprocal. A more complex type also contains the less complex types. The order of complexity is also the order of most difficulty of integration. Integration (Lawrence & Lorsch 1967) can be defined in construction terms as the quality of the state of collaboration that exists among contributors to projects who offer specialist skills and who are required to achieve unity of effort by the environment. If, therefore, there are different types of interdependency, there would need to be different methods of integration. As reciprocal interdependency is the most difficult to integrate, and as this type of interdependency dominates in the construction process, the integrative mechanisms and effort need to be of a high order.

It has been found that the integration of pooled interdependency is best achieved through standardisation and formal rules, and sequential interdependency through planning. That is, the tasks to be undertaken can be anticipated and their sequence planned so that sequential interdependency is identified and recognised at an early stage. The managing process should then ensure that integration takes place as planned. Reciprocal interdependency is integrated by mutual adjustment and feedback between the parts involved.

The integration of reciprocal interdependency requires close association between interdependent parts to ensure that the required input takes place at the appropriate time and that account is taken of the various inputs in the process. The management of reciprocal interdependencies requires that a balance be maintained between the inputs in terms of clients' objectives. A clear perception of clients' objectives is required, together with the diplomacy and expertise necessary to integrate a group of highly skilled professionals.

A study by Mohsini and Davidson (1992) attempted to measure the impact of some organisational variables on the performance of projects which used the traditional building process. They found that two factors, the sufficiency of starting information which stresses the importance of information exchange and the extent of the tasks' interdependence, were consistently important. Interdependency is, of course, generated by the differentiation needed to cope with task and environmental complexity. Whilst recognising that structures without a central integrating mechanism were not suitable for large and complex building projects, they also suggested that newer procurement strategies should seek to reduce the negative impact of 'task interdependence'. They suggest that the use of process-related innovations which reduce the number of participants would be of benefit, for example design-and-build and reducing on-site operations using prefabrication. However, whilst these are useful suggestions, the opportunity to reduce interdependency is limited as it is determined predominantly by the nature of the task and the impact of the environment in the project, both of which are tending to become more complex.

In an organisation which is strongly differentiated yet largely interdependent, such as that found for construction projects, the key to success is the quality and extent of the integrative effort provided by the managers of the organisation. The root of project management should therefore be the integration of the organisation. This applies whatever the organisation structure adopted. Within any organisation there will be someone, or some group, responsible for managing the process. Conventionally, the architect both designed and managed. Increasingly, a project manager is appointed to manage the process. In either of these cases, or any other, the manager's fundamental activity is integration.

The mainstream literature on organisation still finds the use of task forces and project teams as 'complex and fascinating' (Scott 1992). The predominant concern is often their relationship with the hierarchy, the high differentiation within specialist project teams and the need for strong integrative efforts which are seen as exceptional arrangements. These views are also reflected in general management's disquiet with matrix management structures. This is opposite to construction project management for which such structures are *de rigueur* and hence second nature to professionals in the industry. For these reasons the general management literature requires reinterpreting to distil that which is useful for project team organisation and management.

Integration will be necessary in two specific directions. One is the integration of the people involved with each specific task. At a basic level the manager will need to ensure that the appropriate people with the right skills are involved at the correct time. This may seem obvious but it is surprising how often this does not happen. If advice is given too late in the process, or if it is not given at all or not taken, it can lead to abortive work or delay and dissatisfaction by the client with the outcome of the project. So within each task the integration of the contributors needs to be ensured. This type of integration is achieved at a personal level through the characteristics of the manager. The manager should create in the project team recognition and respect for the contribution of others by all members so that the member responsible for carrying out a specific task automatically seeks advice. The manager should then monitor progress within tasks to ensure that the development of the project reflects the level of response between contributors considered to be necessary.

The other, and equally important, need for integration is between the output of the tasks. Each task undertaken by the project team will have to be compatible with each other and in relation to the project's objectives. The manager of the process, who should be taking an overview of the various tasks being undertaken, should evaluate the output of the tasks in terms of their compatibility. This will require what is in effect a feedback loop at each output at which the latter is 'measured' against the project objectives and against the output of other tasks. For instance, it is unacceptable for the design of the electrical services to satisfy the client's objectives for power and lighting if the proposed solution cannot be provided within the cost limit for the work. Adjustment would have to be made to the design, the cost limit, or the client's objectives. Sound integration within tasks should be designed to prevent such an occurrence, but integration of the output (between tasks) acts as a 'backstop' with formal feedback opportunities to ensure that within-task integration has taken place effectively.

Empirical evidence of the relationship between integration and project performance is provided by Pocock *et al.* (1997). Integration was represented by the degree of interaction among designers, builders and project team members during all project phases. Analysis of 38 projects showed significant correlation between integration and performance, including user satisfaction. Beyond a certain level of integration, performance tended to level off and achieving that level took only a modest increase in integration. However, the study took no account of the environments of the projects and hence the relationship between the integration needs of the projects and their environments.

The integrative mechanisms designed into the organisation structure will depend on the particular project and its environment, but will range from integration through personality to formal and rigorous feedback mechanisms at key points. This is understandable as differentiation is present within the system for two main reasons. One of these is to do with the emotional orientation or characteristics of the people involved in the process and the other is related to the technical nature of the projects themselves, which are often complex and demand input from a number of skills to tasks that have to be combined to produce the completed project.

Whilst the construction industry has wrestled with the demands of working in teams since the first building was erected, it appears that mainstream management has only recently recognised team working (*Manager Update* 1997). A major review of the effectiveness of teams by Cohen and Bailey (1997) identified four types of team; work teams, parallel teams, project teams and management teams. The findings for both project and management teams have relevance for construction project teams. In contrast to other teams, autonomy was not associated with higher performance. The quality of leadership was most important for project team effectiveness. For management teams, social integration and shared beliefs were of high significance for their effectiveness. The need for a clear goal for teams and their commitment to the goal, as conditions contributing substantially to success, came through clearly in the review. The work of Larson and LaFasto (1989), which examined projects such as the space shuttle Challenger investigation team and McDonalds Chicken Nugget team found that 'while the team goal was always there in the effectively functioning teams, with almost as much consistency, it was missing in the ineffectively functioning teams'. All of these are findings that resonate with members of construction project teams.

Decisions and their effect on structure

As referred to in the last chapter, the decision-making process and specifically the timing of decisions made by the client can have a significant influence on the effectiveness of the project organisation. The decisions the client makes will be based upon recommendations or alternatives presented by the project team, usually through its manager. Such decision points act as 'pinch points' through which the project must pass if it is to progress. If an acceptable decision cannot be made, the project will not squeeze through the 'pinch points' and will terminate, or the objectives will be revised. Between these major decision points will be others at which decisions will not normally be made by the client but by the project team manager, depending upon the authority pattern that has been established.

Decisions taken at the project team level will contribute to those taken by the client and at each level the manager of the project team needs to integrate the various tasks to produce the alternative propositions available. The manager will then make the decision when it is within his or her authority or make a recommendation to the client if the latter is to make the decision. In order to stand the best chance of making the correct decision, the range of available alternatives and supporting arguments will have to be presented in each case. It is the manager's role to make sure that all the alternatives are exposed and to achieve this he or she will have to take account of the advice of all the contributors to the project. Integration of the contributors therefore assumes paramount importance. The integration of contributors within and between tasks is important, but the key integrating activity of the manager is in bringing together the output of the tasks in a way that allows the range of available alternatives to be clearly exposed.

Thus the essential determinant of the structure of an organisation for the design and construction of a project is the arrangement of decision points and the way in which the contributors need to be integrated in order to produce the material upon which decisions can be made. The most significant decisions are taken by the client and the timing and sequence of the decision points will be determined by the internal organisation and external environment of the client's organisation. The framework needs to be elicited by the manager of the project team from the client before the positions of the decision points for decisions within his authority can be determined. The manager will then be in a position to design the integration of the contributors in the project organisation for the purpose of both classes of decision. Whereas it may be possible to identify a list of routine decisions that are common to all construction projects, it is not possible to determine when they will need to be taken until the framework of the client's decision points is known. Nor is it possible to identify non-routine decisions within the authority of the manager until the client's decision framework is established. Given knowledge of that and of its level of certainty, managers are able to integrate decisions and the contribution of their team in the most effective manner.

A major task of the manager of the project will therefore be to make the client realise the fundamental nature of its role in the construction process and the way in which it can affect the effectiveness of the process and the client's own satisfaction with the project outcome.

Differentiation and integration in practice

Differentiation and attempts to integrate are expressed in a whole variety of ways in practice. Experience on one project with a particular group of participants is not necessarily transferable, either to another project or to another group of participants. Each project should be analysed individually to identify the type and scale of differentiation as a basis for designing the appropriate organisation structure and integrating mechanism. Later chapters will examine how this may be achieved but at this point it may be useful to look at some of the problems and solutions which have been commonly tried in practice.

The most positive approach has been the creation of multidisciplinary practices that employ within the one firm all the professional skills associated with projects. If, within such practices, specialists work in project-dedicated teams, then one would expect that conditions would be created in which a high level of integration could occur. However, if such practices continue to organise in 'departments' of specialist skills, a great integrating opportunity will have been lost. In either case, integration with the contractor will be difficult to achieve if the contractor is not appointed until after design has been substantially completed. Even if appointed early in the process, differentiation will be high and special integrating effort will be required. A similar situation will exist on a development project if, as often occurs, a general practice (valuation) surveyor is involved from outside the consortium.

Differentiation is high on a project when professional consultants are from separate firms and they will be differentiated from the contractor to varying degrees depending upon when and how they are appointed. If positive attempts are not made to integrate them, the effect upon the project outcome can be serious. An example encountered was a project for which the quantity surveyor and services engineer never met but communicated indirectly. The result was that the services installation cost control was badly managed, leading to abortive work and dissatisfaction by the client. A more positive approach was discovered on another project on which the professional consultants, although from separate practices, worked together for the whole of the design phase in the office of the consulting engineer who also provided the project management service. At first the consultants did not like the idea of uprooting themselves and working in unfamiliar surroundings but after the event they agreed that it had been very beneficial in creating a harmonious team and producing a project that met the client's requirements.

The familiarity of the contributors with each other's methods as a result of working together on previous projects does, of course, assist integration but can lead to complacency. This can be evident where the same team works on a subsequent project, which places different demands upon the team as a result of the client's requirements and the environmental influences acting upon the process. There is a danger that there may be no stimulant within the team for a change of approach.

An interesting paper by Baiden *et al.* (2006) examined the extent of integration within nine construction projects and found that the projects exhibited varying degrees of integration determined by the team practices adopted and their congruence with the procurement approach. Design-and-build had the highest integration, construction management had good integration and traditionally procured projects showed some evidence of integration, all as would be expected. No team was 'seamlessly integrated', but this was found not to be a fundamental requirement of integrated team performance. However, the research did not start from the basis of systems theory, in particular contingency theory, as the nature of the task, i.e. the nature of the project, was not taken into account and data was based on interviews (with award-winning project managers). Thus there was no attempt to measure the degree of integration required of each project as a function of the extent of the differentiation generated by the project.

A frequently occurring event having a great impact on the quality of integration, and which has been consistently overlooked, is the highly disruptive loss of key project personnel. Chapman (1999), in aiming to correct this oversight, showed that on 22 projects examined, 'the most frequently occurring risk within the design stage (out of the 13 most common risks identified) was changes in key personnel'. The projects, which had severe programme overruns, showed a clear link with changes in key staff.

A later study (Eskrod & Blichfeldt 2005) found that the 'project manager should acknowledge that team composition is not static' and that 'the project manager should be conscious as regards managerial challenges associated with each member's enrolment, entrance, integration, maintenance, and withdrawal'. They 'suggest that formal transition and maintenance rites are introduced and, in particular, that a formal mentor be appointed to help entrees according to their needs'.

The extent to which trust exists between team members is something that has received little attention, although Munns (1995) has addressed it. He defined trust as 'a decision to become vulnerable to or dependent on another in return for the possibility of a shared positive outcome'. It can therefore be seen to be a powerful psychological force in team integration. Personnel employed temporarily on projects were seen to view the development of long-term stable relationships as a secondary concern. On the basis of the arguments he presents, he concludes that there are two key factors in determining the relationships which will exist on a project:

'these are the levels of global trust experienced by all project members and the early actions of the team members in the formation of the team. It has been suggested that the first actions of the team members will be dictated by their level of global trust and it is the shaping of this global trust that needs to be understood.'

It is concluded that the initial opinions of individuals of their co-team members on entering the project are important in shaping its final outcome, as they can force the project into a spiral of increasing or decreasing trust. Explanations of the levels of trust exhibited can lie in psychology and transaction cost economics although the paper does not address these issues. These ideas also make a contribution to the development of partnering as discussed later.

In a complementary paper, Munns (1996) analysed the mutual confidence between clients, contractors and professionals. He found that,

'the high level of confidence afforded to professional bodies indicates that both clients and contractors consider these to have a high level of integrity. The lower levels of trust shown towards both clients and contractors indicate a potential source of conflict. If the trait characteristics of low trust are perpetuated within a specific contract, the potential for a breakdown in relations is high.'

Again these findings have implications for the widespread use of partnering approaches.

As mentioned previously, the problem of integrating the contractor is always present. In examining occupational stereotypes in the construction industry, Loosemore and Tan (2000) found that the relationship of the contractor with the architect and quantity surveyor was more problematic than other relationships, 'because stereotypes held between these occupations are relatively strong and negative'. Many of the new initiatives of contractor involvement in the design team are intended to try to assist the integration of the contractor into the project team. Design-and-build, management contracting and the construction management approach are examples of initiatives that bring the designing and constructing functions closer together. Although the benefit claimed for these approaches is that they allow an input of construction knowledge to the design, there is a potential for an equally important benefit in terms of integrating the people working on the project.

Design-and-build has emerged over recent years as an increasingly popular procurement method. Its appeal to clients arises from single point responsibility which simplifies the manner in which the client interacts with the project team. Theoretically a design-and-build organisation structure should reduce differentiation and provide a sound platform for effective integration resulting in a proficient management structure. However, in practice this may only be the case if all the project skills are in-house to the design-and-build company. If this is the case differentiation will be reduced and an in-house project management culture can be developed leading to high integration. But frequently design-and-build companies do not have all professional skills in-house. There are a number of reasons for this, including the desire to limit investment and the problem of retaining the scope of skills (particularly design) which may be required. As a result professional skills are hired in from individual professional practices. The design-and-build company in such circumstances, which will be based on contracting, will have the responsibility for integrating the project team. The organisational advantage in such circumstances is questionable although the client still has the advantage of single point responsibility. The design-and-build company will need highly developed project management skills as there may be a reluctance on the part of members of professional practices to be managed by construction companies unless the practices are carefully selected. The client may also retain a project manager and other professional advisors to oversee the design-and-build company. If professional skills are in-house to the design-and-build company the relationship with the client's advisors should not be too difficult to manage but if the design-and-build company hires in professional advisors the situation becomes much more complex due to the multiplicity of firms involved.

The appointment of a project manager on behalf of the client, either in-house to the client organisation or from an independent firm, should act basically as an integrating device although the benefit of such an appointment is often justified in other terms, such as progress chaser, controller, etc. or even just as a preoccupation with titles. One would expect that the greater the differentiation between the contributors, the greater the need for a project manager. The latent differentiation of contributors to all projects as a result of sentience and other forces means that the need for integrating effort is always high and would probably benefit from the integrative effort provided by someone solely concerned with project management.

Most building projects require someone to act as a catalyst. This need is often recognised for large complex projects and there is no doubt that the scale and complexity of such projects, both technically and environmentally, expose differentiation and demand an integrating mechanism such as a project manager. However, the extent of differentiation on the medium-sized or even the smaller projects is not as readily recognised but is sufficient to require positive action to integrate the contributors rather than just hoping that it will happen.

International projects can generate the greatest levels of differentiation. Not only will the contributors be differentiated for the reasons given previously, but such differentiation will be compounded by the contributors coming from a variety of countries and being required to apply their skills in a country with which they may not be very familiar. The differences in cultural background and methods of working will generate differentiation, which can only be integrated by careful organisation design and very positive effort. It is therefore not surprising that project managers are employed more frequently on international projects than on national ones. This need is also reinforced on overseas projects by the environmental and often technical complexity of the work, as projects are frequently undertaken to build whole industries and extensive facilities from scratch in conditions of great uncertainty.

Theoretically, to reduce differentiation to a minimum, clients should develop their projects using a team of specialist skills as employees within their own organisation ('in-house') including the construction phase using directly employed labour. In such a situation the likelihood of conflicting objectives among the contributors would be reduced. The allegiance of the contributors should be largely to the client directly, although allegiance to professional skills would not be eliminated altogether. In this type of arrangement the opportunity to generate the maximum level of integration using an in-house project manager, who would have access to and a full understanding of the organisation's objectives, should be at its greatest. However, in most cases this is just not practicable as most clients do not have such an in-house capability. Local authority direct labour organisations in the UK were one of the few examples of this type of arrangement but their history is not generally one of great success, although there are some successful examples. Perhaps their problem was more political than management orientated. Nevertheless, the in-house scenario is an interesting one upon which to base thinking about the type and degree of differentiation and integration present in any specific project organisation.

Whilst the preoccupation here has been on the practical issues of integration, it is important to remind ourselves of the fundamental theoretical basis of the differentiation in need of integration. This lies in transaction cost economics as discussed in Chapter 3. The degree to which the components of the project team are in-house or outsourced (and hence more differentiated) is a function of the transaction costs incurred by alternative arrangements, (i.e. market or hierarchy) other costs being equal. In practice the precise cost of each alternative may not be calculated and great reliance may be placed on intuition, nevertheless the underlying criteria will be the costs of the alternatives. However, a short-term perspective of cost may be taken, which is inappropriate. If the effectiveness of the project in the long term is not taken into account, the choice of project structure may not be optimal.

O'Brien *et al.* (1995) make the point that the predominant use of the market mechanism for co-ordination of firms for project execution limits the potential for Information Technology (IT) to add to value at the project level. Whilst they were considering main contractors and subcontractors, the same arguments apply to project teams. They argue for centralisation (hierarchies), which they claim requires information technologies and hence, provides an economic rationale for the adoption of supporting IT. This appears tautological. The fundamental criteria are economic. In the long run transaction costs will determine whether organisation structures are based on markets or hierarchies (or an appropriate amalgam) and this will incorporate the costs (and benefits) of IT.

Organisational culture

Sentience and groupthink, which have been discussed previously, can be seen to be specific aspects of organisational culture that impact on construction project management. Organisational culture is a broader and fuzzier concept, as demonstrated by its many and varied definitions (Palmer & Hardy 2000; Fellows 2006). Its clarity is further clouded by the relationship between organisational climate and organisational culture, which Westwood and Clegg (2003) see has 'never been an easy or clearly delineated relationship'. Put simply, organisational climate is seen as 'the beliefs, and attitudes held by individuals about their organisation'

(Brown 1998), whereas organisational culture has been defined as 'the collective programming of the mind which distinguishes the members of one organisation from another' (Hofstede 1994). In summarising a debate on 'Making sense of culture', Westwood and Clegg state 'the protagonists are in agreement: "climate" has a clearer construct development, delineation and consensus, "culture" on the other hand, is in relative disarray'.

Adding to the confusion are the numerous ways which have been suggested for classifying culture. For example: role, task, power and person culture (Handy 1985 in Fellows 2006), social, technical and managerial culture (Naoum 2001), and hierarchy, market, clan and adhocracy (Cameron & Quinn 1999). Not helping the application of organisational culture ideas to construction is the notion that 'organisational culture may be thought of as a reformulation of existing models and theories to satisfy changing views of how organisations work, and which is likely to enjoy its pre-eminence for a finite period of time' (Brown 1998). This view is supported when a few years later Palmer and Hardy (2000) state: 'One debate here is whether culture has had its fifteen minutes of fame,' 'Indeed some have even proclaimed the death of culture,' and 'One view attributes the death of culture to its limited practical relevance – so many attempts to change it and improve organisational performance have failed'.

Construction project management studies would therefore appear to have come to culture at a late stage in its life cycle. Nevertheless, there are benefits from understanding the impact of culture on the process of managing construction projects. Much of the mainstream management literature is about how changing an organisation's culture can lead to improved organisational performance. This is not necessarily the focus which is most useful for managing construction projects. Managing construction projects is about harnessing the various cultural forces at play on construction projects to the benefit of projects, and hence clients, in meeting their objectives. There is an argument that the design, development and construction of projects need contributions from different organisational cultures, therefore project managers need to understand the different cultures present in projects so that they can harness them and not fight them.

Scott (1998) states that most analysts focusing on culture embrace a functionalist perspective and concentrate on the contributions that cultural elements make to organisational unification and control. This leads to the view that culture can be managed. Contrary to this is the view that culture is a process of natural change, that new members 'acquire' culture through socialisation and it is not directly manageable (Ogbonna & Harris 1998). However, the conception which appears most relevant to construction organisation, is the functionalist assumption as espoused by Schein (1992) that, 'These assumptions and beliefs are learned responses to a group's problems of survival in its external environment and its problems of internal integration'.

Schein (1996) also believes that there are subcultures at work and defines what he calls the 'operators' group, the 'engineers' and the 'executives'. He believes they present a built-in conflict between subcultures and that they have learned their assumptions from the general occupational environment within which they exist. Thus their responses to environmental forces may not reflect a consistent organisational culture. Hopkins *et al.* (2005) identify technical and non-technical subcultures in technology-intensive organisations. They conclude that managers of such organisations must identify the core values 'and then make sure they are clearly understood and accepted by both technical and non-technical workers'. Although these classifications do not relate directly to construction, some of the same forces are at play in the clash of cultures between professions. It is a function of project management to align contributors to the project's core values, as well illustrated by Ankrah and Langford (2005) for the cultural orientation of architects and contractors. Hopkins *et al.* also believe that 'subcultures are often viewed by management as representing tolerated deviations that do not disrupt the normative solidarity of the overall corporate culture's dominant values'. This view highlights the difference between managing intra-firm subcultures and the inter-firm subcultures prevalent in the construction industry.

Scott (1998) refers to the idea in the popular organisation literature of the benefits of creating a 'strong' culture, which aims to sustain a 'commitment to something larger than self', but warns of this developing into an authoritarian system that is subject to abuse. Newcombe (1997) gives the advantages of a strong culture in construction projects as enhanced effectiveness, that contributors learn to live together and are less parochial, and that it reconciles conflicts. He sees the limitations of a strong culture as difficulty in introducing change, conservatism inhibiting innovation, and isolation of a project from its environment. He implicitly adopts a functionalist perspective by stating that 'culture is built slowly by committed leaders with a strong vision or mission for the project'.

Culture appears to be a powerful force in the construction industry and is particularly complex on projects. The overriding image of the industry is macho, uncompromising, uncaring, opportunistic and adversarial to which Green (1998) adds a culture of 'control and command'. Although some construction firms and their workers may have these cultural traits, they are not necessarily common to all. Certainly not all organisations contributing to projects will demonstrate these traits but therein lies the complexity of the culture of project organisations. Architects and other designers may well exhibit a predominantly aesthetic culture, which may be seen by other project team members as 'airy fairy', engineers may have a culture of inflexibility and quantity surveyors one of pedantry and conservatism. The culture of client organisations complicates this scenario, as their culture is unknown to the project team members at the outset of the project. These are no more than illustrations but the point is that each group of specialist contributors to projects will bring their own culture to the table even if all contributors are in-house to the client organisation. The differences are likely to be greater if they are from separate organisations. There appears to be a belief that all contributors have an adversarial component in their cultural profile. Thus we find that the ideas of culture in organisations directly reflect and restate the ideas inherent in the concept of differentiation discussed in detail earlier. Such adversarial attitudes raise the issue of the relationship between culture and trust in the construction industry. Generally, it could be said that mistrust is a major traditional position between clients and contractors and between consultants and contractors, and it is discussed in more detail later in the context of opportunism in transaction cost economics. Efforts to generate trust have gained momentum with the advent of partnering and other relational initiatives, which seek to change the culture of the industry through arrangements for specific projects.

A major contributor to the culture of the professional contributors to projects is professional ethics which, in this context, are essentially the rules of conduct recognised in a particular profession. Whilst many members of the professions may see their ethical responsibilities as much wider than the rules of conduct, those rules do form the bottom line. Professional ethics are one element which distinguishes construction organisations from other commercial organisations and hence is a distinguishing feature of their cultures. Fellows (2006) gives a good account of ethics and believes that they 'constitute a vital behavioural link between culture, climate and behaviour of members of an organisation'.

Cultural differences illustrate the difficulty of the task of the project manager in integrating project teams and raise the question of whether it is possible to develop a project culture that is constructive in producing effective project teams. The acceptance of a functional view of culture implies that culture can be managed. That is, a project manager, given appropriate leadership attributes is able to develop an appropriate culture within a project organisation. The type and duration of a project are likely to have a significant effect on the opportunity for a project manager to develop a project culture. A high management profile may give the project manager the chance to imbue a strong sense of purpose in the project team and the longer the project duration the more time the project manager will have to work on cultural aspects. A major attribute would be charisma and the consequential respect of the project team but, even given the best talents, they would need to be sufficient for the project manager to overcome the forces working against the establishment of a project culture. Loosemore and Tan (2000) in their work on stereotypes in the construction industry remark that breaking habitual prejudices 'involves changing a lifetime of socialisation and experiences and can be achieved only by exposing people to their stereotypes and prejudicial behaviour and by showing them how to modify their behaviour in a non-prejudicial fashion'.

A further idea which qualifies the functional view of culture is that norms tacitly accepted across wider social units such as entire organisations or occupations are much more likely to change leaders than to be changed by them (Schein 1996). Whilst this may not have an effect on project organisations due to their relatively short-term temporary nature, it could certainly be a force in defining the culture of the contributing firms. Project managers also need to guard against a 'gap between what management espouses and that which the organisational members believe to be true' (Ogbonna & Harris 1998). In arguing against the adoption of business process re-engineering in construction, Green (1998) states that, 'The notion of corporate culture is less about shared values and more about "brainwashing" individuals into accepting the required mode of behaviour'. Partnering is an approach that seeks to address adverse cultural problems, particularly the destructive nature of the adversarial attitudes on project team members (Liu & Fellows 2001).

Whilst the focus here has been on organisational culture, culture has different levels, for example industry and global levels (Palmer & Hardy 2000). These different levels can impact on the organisational level, as shown by Rowlinson (2001). In investigating the implementation of a matrix management structure in a construction-based professional department of the Hong Kong government, he found that what 'appeared to be a mismatch between organisational culture as perceived by the workers and the organisation structure that was being implemented' was in reality due to issues which could be 'attributed to deep-seated, traditional cultural values' of which the Chinese concept of 'face' was a major part.

In many ways, the concept of culture can be seen to be no more that the consolidation and restatement of that which has gone before in the study of the behavioural aspects of organisations. Consequently, Ogbonna and Harris (1998) state, 'However, whilst a conceptual understanding of the intricacies of the managing culture debate has provided many worthwhile theoretical contributions, many theorists have noted the need for further empirical work in this area'. The same view is also appropriate to its application to construction.

Partnering

Partnering emerged in the late 1980s as a form of collaboration between companies within construction project organisations, particularly between clients and contractors but also between contractors and subcontractors. One of its major aims is to improve co-operation and reduce the adversarial nature of the industry. Its focus is behavioural rather than structural and, in terms of systems theory, can be seen to be directed at reducing differentiation and increasing integration. Partnering has been subject to a plethora of definitions (Li *et al.* 2000) but an early one by the US Construction Institute (1989) remains convincing:

> 'A long-term commitment between two or more organisations for the purpose of achieving specific business objectives by maximizing the effectiveness of each participant's resources. This requires changing traditional relationships to a shared culture without regard for organisational boundaries. The relationship is based on trust, dedication to common goals, and understanding of each other's individual expectations and values. Expected benefits include improved efficiency and cost effectiveness, increased opportunity for innovation, and continuous improvement of quality products and services.'

Whilst the implication of this definition is for an ongoing relationship over many projects (strategic partnering), many other forms have appeared, such as partnering on individual projects and 'partnering' with various levels of co-operation many of which fall short of full partnering but still carry the same name (Matthews *et al.* 1996).

Practically all the research and development in partnering in construction has been concerned with techniques and application. It has been mainly prescriptive and uncritical. Bresnen and Marshall (2000a) draw attention to this failing. They point out that even though partnering is held up to be about changing attitudes, improving relationships and developing a partnering culture, little research has explored the social and psychological aspects of partnering, to which can be added economic aspects. Green (1999) also reflects on the uncritical acceptance by industry of partnering. He believes that, 'The buying power of the industry's major clients continually discourages dissent to the partnering ideal. Construction companies which do not appear similarly committed risk being denied access to a substantial part of the UK market'.

To fully understand partnering in all its forms, it is necessary to seek out the underlying reasons for its development in order to deduce the theory on which it is based. This has its roots in transaction cost economics, trust, risk and culture all of which underpin the characteristics in the definition above. Their relationship has been addressed in the mainstream management literature but not in relation to construction. For example, Chiles and McMackin (1996) argue that what they perceive as, 'shortcomings of the transaction cost economics paradigm are in part attributable to the inadequate treatment of risk and trust'. For partnering to be more cost effective than other forms of relationship it must reduce production and/or transaction costs. Any reduction in production costs, which results from more effective ways of working as a result of partnering are funded by the transaction costs of partnering. The economic argument is, if the sum of production and transaction costs is less for partnering than alternate forms of agreement then partnering will be adopted. Thus the significant element is transaction costs, which comprises the costs of entering a partnership arrangement.

The major aspect of transaction cost economics having relevance here is the assumption of opportunism. Opportunism essentially means, 'self-interest seeking with guile' and has profound implications on the choice and design of appropriate governance structures and organisational arrangements (Williamson 1975). Transaction cost economics argues that firms will seek to protect themselves against opportunism, often by contractual arrangements. However, partnering adopts a fundamentally different approach by basing the relationship between the parties on trust as a protection against opportunistic behaviour. In its conceptual form, partnering should not require contractual agreements between the parties but in reality they are invariably used but with the expectation that they will not have to be relied upon. The expectation is that partnering will reduce transaction costs through more amicable arrangements between the parties that will reduce the costs of protecting against opportunism and also, as an additional benefit, by reducing production costs through, for example, joint innovation.

In contrast, for traditionally arranged construction projects, firms enter costly contractual arrangements. The costs of such arrangements are part of the transaction costs of producing a building. Williamson (1985) recognises such a situation and states, '... parties to a bilateral trade create credible commitments, whereby each will have confidence in trading with the other'. Credible commitments are seen as contractual safeguards, assurances and mechanisms. He goes on to recognise the differential hazards of breach that arise under different investment and contracting scenarios and presumes that suppliers are far-sighted and will bid for any contract for which an expected non-loss-making result can be anticipated and that buyers will choose the contractual terms that best suit their needs. The parties therefore have a mutual interest in devising an exchange relationship in which both have confidence (Yates 1998). Partnering is one such exchange relationship but one in which trust forms a significant part of the credible commitment. However, transaction cost economists find what Williamson (1990) describes as 'the use by social scientists of user-friendly terms such as trust' to be dubious as credible commitments. He states,

> 'The growing tendency to use trust to describe probabilistic events from which the expected net gains from co-operation are perceived to be positive seems to me to be inadvisable. Not only does the use of familiar terms (like trust) invite us to draw mistaken parallels between personal and commercial experience, but also user-friendly terms do not encourage us to examine the deep structure of organisation. Rather, we need to understand when credible commitments add value and how to create them, when reputation effects work well, when poorly and why. Trust glosses over rather than helps unpack the relevant micro-analytic features and mechanisms.'

Yet he does not dismiss trust, but distinguishes between personal and commercial trust and asks for depth in its analysis. In support of trust he cites membership of professions that are self-regulated as giving trading confidence to transactions and also recognises that, for the purpose of economic organisation, culture serves as a check on opportunism. Nevertheless he does say that 'transaction cost economics refers to contractual safeguards, or their absence, rather than trust, or its absence'.

Such theoretical attention that partnering has received has focused on trust and culture without recognising the relationship with transaction costs. Essentially, partnering is designed to reduce transaction costs by, for example, greater collaboration for greater efficiency, not incurring cost associated with withholding information (through co-operation), elimination of adversarial attitudes and hence the costs of disputes. In order for such things to happen it is claimed that a change in culture within the partnering organisations has to be engendered. That is, a change in organisational culture is needed in order to develop the trust necessary to make partnering work effectively. But, as Williamson argues, trust has a cost – a transaction cost. The cost of trust is the cost of the risk of trusting your partner without credible commitments, e.g. contractual safeguards, to protect against the performance expected of your partner not being achieved. Thus organisational trust and culture are important in understanding and developing partnering but only in the context of a reduction in the costs of projects, which must be broadly defined to include production costs, transaction costs and the life-cycle costs of the project to the client. The risk of partnering lies in one of the partnership not fulfilling their part of the bargain with no redress available to the other party. The costs incurred by such a breach of trust would generate increased transaction costs which could result in the combined production and transaction costs being greater than would have been incurred using different arrangements, e.g. more formal contractual arrangements. Thus part of the transaction costs of partnering is undefined as it is embedded in this risk.

Trust is rarely mentioned in the literature relating to partnering in construction. Rather, reference is made to the need for cultural change, which improves co-operation and reduces conflict. However the underlying characteristic needed for such changes to take place is trust between the partners. The mainstream organisational studies literature does not ignore trust but recognises it as a concept central to alliances and partnerships. Mayer *et al.* (1995) link trust and risk by drawing on Johnson-George and Swap (1982) who asserted that 'willingness to take risks may be one of the few characteristics common to all trust situations' in their definition of trust:

> 'The willingness of a party to be vulnerable to the actions of another party based on the expectation that the other will perform a particular action important to the trustor, irrespective of the ability to monitor or control that other party.'

They distinguish trust from co-operation, confidence and predictability, which they claim have often been confused with trust. The overriding characteristic of trust is seen to be vulnerability and the risks associated with it. Partnering in construction tends not to extend so far because contractual safeguards are also used. The objective of partnering is focused on developing a culture of co-operation rather than trust. As they state, 'Although trust can lead to co-operative behaviour, trust is not a necessary condition for co-operation to occur'.

That trust is not well understood is illustrated by contrasting the absolute view of trust above with that of Das and Teng (1998) who examine it in the context of alliances of two or more partner firms. They see trust as a source of co-operation and confidence between partner firms and define partner co-operation as 'the willingness of a partner firm to pursue mutually compatible interests in the alliance rather than act opportunistically' and confidence as 'a firm's perceived certainty about satisfactory partner co-operation'. They argue that confidence in partner co-operation comes from both trust and control. Control is seen to be 'a regulatory process by which the elements of a system are made more predictable'. They use joint ventures (JV) as an example in which the partner firms regulate what the JV does and also what the partners may do. Trust and control are seen as parallel concepts and their relationship of a supplementary character in generating confidence. Hence there may be different levels of trust in an alliance and different levels of control. This conception is intuitively more acceptable to construction partnering. This could stem from the client–contractor relationship being one of buyer–seller rather than the relationships in the management literature, which tend to be of two or more firms forming alliances to produce goods or services to sell to others. The former may inhibit the extent of the development of trust and create the need for controls.

The issue of trust in construction partnering is not often addressed from a theoretical standpoint. The difficulty of dealing with it is illustrated by Lazar (2000) in which he concludes that, 'trust based relationships are . . . critical to maximizing positive economic outcomes from partnering and may be necessary to keep the owner/contractor relationship from deteriorating – even those based on mixed (co-operative) strategies'. Yet a mixed strategy comprises collaborative and competitive strategies. He also points out that causes other than trust, for example fear or coercion, may lead to collaboration. The role and contribution of trust in partnering is therefore far from clear when there is scope for opportunism and the need for contractual arrangements to protect against it. Wood and McDermott (1999), in attempting to relate trust to partnering, demonstrate the elusiveness of the concept. They also recognise that there are degrees of levels of trust (which, although they do not make it clear, will need credible commitments to complement the level of trust). Others treat trust merely as a factor of success rather than an explanation of success (Li *et al.* 2000). Whilst Lazar's arguments are essentially uncritical of partnering, he does point out the most fundamental need for empirical testing of theory. But maybe theory building is needed first.

The earlier definition of partnering states that, to be successful, members need a shared culture based on trust. So the key is a shared culture. The earlier reference to culture identifies the conventionally accepted view of the culture of the construction industry as one of animosity. Partnering seeks to change this to one of co-operation, collaboration and above all trust. There are obviously severe problems in achieving this, even when the partnering relationships are intended to extend over a long period and many projects. A shared culture is much more difficult to achieve on a single project, even when it is of long duration. Nevertheless there is much anecdotal evidence of successful partnering for which a shared culture based on trust has been achieved (Liu & Fellows 2001).

The functionalist perspective of culture believes that culture can be managed. Partnering implicitly accepts that this view, as the alternative of a naturally emerging culture that is not directly manageable, is unlikely to have the chance to mature on anything other than extremely long-duration partnerships. Such relationships are unusual in construction/client relationships but could occur more frequently between main contractors and subcontractors. The management of the large cultural shift needed to achieve trust in partnering places huge demands on the managers of the process. Charismatic leadership and a powerful commitment by the leaders of all parties involved are necessary. Even with such commitment at the higher levels of management, it is vital to ensure that it percolates to all levels of the organisation otherwise high ideals could be undermined at the operational level. Hence the calibre and commitment of project managers on partnering projects need to be of a high order.

As Bresnen and Marshall (2000a) state, 'partnering is a rather loose term to describe what is in reality a multi-faceted practice, but clearly also it is not always seen as necessary or desirable'. They argue convincingly for what is essentially a contingency approach to partnering based on sensitivity to local conditions. They recognise, of course, that partnering is not an organisation structure but a manifestation of organisation culture, which can be cultivated in whatever organisation structure is appropriate to the project. It is important to recognise that partnering does not eliminate the need to structure the project organisation effectively. Whilst partnering can be seen to be about the integration and compatibility of the objectives of the firms making up the partnership, at the end of the day the success of the project will depend to a large extent on the integration of the people working on the project, as is the case with all organisational forms. The key is whether partnering allows this to happen more effectively.

Larson (1995) found that it did. He studied 280 construction projects in the USA to identify the relationship between project success and alternative approaches to managing the owner–contractor relationship, of which one was partnering. The results showed that partnered projects achieved superior results in controlling costs, technical performance and in satisfying customers compared with those managed in what was termed adversarial, guarded adversarial and on an informal partnered basis. Whilst he points out that the study was exploratory and has limitations, nevertheless the results are important. He predicted the inferior results for adversarial and guarded adversarial approaches but found the differences between informal and fully-pledged partnering significant and well worth the investment in full partnering.

However, in a rather more detailed study, Bresnen and Marshall (2000b) found that generally whilst there are many potential benefits for clients in more collaborative approaches, nevertheless conventional projects can show the same benefits and that a partnership does not guarantee them. In particular, 'collaborative projects did not necessarily remove conflicts at source and that there was still the persistence of major problems in integrating design and construction' but that there was also 'clear evidence from the research of the avoidance of potential claims and disputes...'. They presented some valuable insights to the complexity inherent in the choice of partnering and found that fully-fledged partnering was not always 'necessary, desirable or feasible,' particularly in cases where there were difficulties in providing continuity of work. Misgivings were also expressed about 'long-term relationships being too "cosy" and uncompetitive'. The efficacy of many of the formal mechanisms used to develop partnering, particularly incentives, was questioned. This latter issue is well developed in detail in their complementary paper (Bresnen & Marshall 2000c). Team building was seen as vital to success and should permeate all levels of the organisation and not reside simply at manager level, hence excluding client level and site level. A further important observation was that, although change efforts had been successful, 'collaboration was still fragile' and that there 'needs to be continued senior management support for collaboration'. Also significant was continuity of relationships, which has been referred to earlier as important on conventional projects. The impact of key personnel leaving on partnered projects is even greater. Whilst supporting partnering generally, this important paper gives a balanced view of the practice of partnering, its 'practical problems, limitations and paradoxes'.

Subsequent studies have indicated that partnering may not be the panacea claimed in some quarters. Whilst in the specific cultural context of Hong Kong, Wong and Cheung's 2004 study found that clients, consultants and contractors relied on what was termed 'systems-based trust', referring to 'legally binding agreements and terms where trust relied on the formalized system rather than personal matters', which appears to contradict the root of partnering. In the

UK, Wood and Ellis (2005) surveyed the commercial managers employed by a leading contractor and found that 'the perceptions and experiences of partnering relationships are generally positive, although the early optimism at the beginning of such arrangements is seldom sustained throughout the project lifecycle'.

Supply chain management

Supply chain management is frequently touted as a panacea of the construction industry's ills, but Cox and Ireland (2002) illustrate the poor thinking within the construction industry in this and other respects. There is certainly a lack of clarity in what is meant by 'partnering' and what is meant by 'supply chain management'.

The terms 'partnering', 'strategic alliances' and 'supply chain management' appear to be used interchangeably (Dainty *et al.* 2001) and with a consequent confusion.

Fundamentally, as stated by Fryer *et al.* (2004), the supply chain is 'all those activities associated with moving goods from the raw materials stage through to the acceptance of the product or service by the end customers'. This process has always existed and has traditionally been managed and achieved through a series of contracts between client, contractor, subcontractors and suppliers, usually based on competitive bids with the lowest bid usually having been accepted. This process has frequently led to adversarial relationships between the parties and poor outcomes to projects. Thus, supply chains have always existed; so why are they receiving greater attention now?

The perceived need to reduce adversarial relationships, which led to the promotion of partnering as a way of improving client/main contractor relations and improving project outcomes, is seen to be capable of being transferred down the supply chain to encompass subcontractors and suppliers. This, together with evidence from manufacturing industries that alliances with other organisations enhanced their performance (Love *et al.* 2002), has led to a surge of interest in how the construction supply chain should be managed. Essentially, the objective is to find new ways of improving integration (a fundamental aspect of systems theory) of the construction process. The Holy Grail of these ideas is that all contributors to a construction project should enter into partnership-style arrangements with all other firms with which they are directly related, both upstream and downstream of them in the supply chain. These arrangements are intended to allow subcontractors and suppliers to contribute to design, programming and other areas of collaboration. They would not normally be appointed by traditional competitive methods, with the objective of building up trust between the parties and removing adversarial relations.

Associated with supply chain management ideas is lean production in manufacturing, which has led to the lean construction movement. London and Kenley (2001) characterise the lean approach in manufacturing (essentially the Japanese motor industry) as improving flexibility, reducing waste and improving flow along the supply chain to provide customers with exactly what they want at the time they want it. They point out that much of the construction literature has applied the lean concept without the detailed empirical exploration of market structures that underpin the construction environment, but also that researchers in the field understand this. Cox and Ireland (2002) believe that 'it is debatable whether such an approach can truly prevail within the current industry structure'. Green and May (2005) are also sceptical about the ideas of lean construction. They conclude: 'The meaning of "leanness" has been found to be characterised by an empirical elusiveness. Yet the essential vagueness of lean construction has not prevented its acceptance as a recommended component of "best practice" ... The continuing popularity of consultant-led training courses is indicative of a mutually beneficial collusion between the promoters and receivers of lean discourse. Consultants need to build constituencies within which they can sell their services, and receiving managers have a continuous need for persuasive scripts against which they can act out the role of improvement champions'.

The limitations imposed by the structure of the construction industry on the applications of these ideas generally has been recognised. Dainty *et al.* (2001) found that the continued reliance on subcontracting for future performance improvement requires an acceptance of the benefits of supply chain partnering and integration from the small- to medium-sized businesses who carry out the majority of construction work. They found serious concerns among such subcontractors due to mistrust and scepticism arising from existing supply chain relationships. The real-world situation is described by Cox and Ireland (2002) as a continuum. The gamut of possible supplier relationships ranges from purely independent transactional, price-based interactions, through highly interdependent relationships, to situations where dependent sourcing arrangements are the only alternative to the organisation purchasing the construction asset. They also point out that because many clients do not understand the industry, they do not know which type of supplier to use, that is, whether to have one supplier which integrates all components of the supply chain or whether to go to the market separately for each component.

The reality is also that there appear to be no published examples of full supply chain integration (Briscoe *et al.* 2004); hence it remains an idealised concept. It is seen to be more likely to be achieved when there is continuity in the supply of projects, otherwise firms are concerned about the financial risk of committing the resources to a single client (London & Kenley 2001). Clients are seen to be the drivers for integrated supply chain management due to their continuity of workload and hence their power to implement such initiatives. Cox and Ireland (2002) term this 'buyer dominance or interdependence where power and influence can be used to force its implementation, rather than those characterised by independence or supplier dominance'.

The discussion so far has been on the practical aspects of supply chain integration; but what are the theoretical explanations? The emphasis on integration of the supply chain stresses once more the fundamental contribution of systems theory to the understanding of the construction process. London and Kenley (2001) 'noted that the development of the idea of the supply chain owes much to the emergence from the 1950s of systems theory and the associated notion of holistic systems'. Although not explicitly recognised, Cox and Ireland (2002) subscribe to the contingency theory of management. When arguing against the universal use of integrated supply chain management, they say that 'while this approach can be made to work successfully in some circumstances, it cannot be made to work successfully in all' and that any corporate 'best practice' is clearly contextually and relationally dependent.

Whilst recognising that systems theory will identify the degree of differentiation required by the project and hence the corresponding level of integration, it does so in an abstract sense in that it does not take into account the cost of integration, which ultimately explains the structure adopted; that is, the degree of collaboration between the components of the supply chain. An explanation is provided, as to be expected, by that which applies to partnering–transaction cost economics, as explored in the last section, because the ideas underpinning supply chain management are an extension of partnering further down the supply chain.

There are several issues determining whether alliances based on early involvement of subcontractors and suppliers making design and other contributions earlier in the supply chain result in a reduction of the sum of production and transaction costs. The element considered here is the transaction costs. As discussed before, the major aspect of transaction costs having relevance here, as with partnering, is the assumption of opportunism and the expectation that alliances will reduce the costs of protecting against opportunism. The discussion of transaction costs relating to alliances under partnering is not repeated here, but the arguments apply equally. Interestingly, the literature on supply chain management makes explicit reference to transaction cost economics only occasionally (London & Kenley 2001), doing so more often obliquely (Cox & Ireland 2002; Dainty *et al.* 2001); but its relevance is clear to see.

Dainty *et al.* make the point that significant barriers remain to supplier integration and to further development of supply chain alliances, even though there is a belief that more integration will solve many of the industry's problems and that those involved wish to be integrated. Past and current 'standard' relationships are thought to be the cause; but transaction cost economics tell us that if additional integration will reduce the sum of production and transaction costs, then it will take place. It would appear then that the perceived costs of protection against opportunism are greater than the perceived reduction in cost which would accrue from greater integration. Mortledge (2006) believes that whilst a partnering relationship where there is a true supply chain makes sense, attempting to transfer this thinking from a product-based culture to a project-based construction scenario is, at best, optimistic (in most cases). He believes that there are few true examples of a supply chain in construction, though supply networks of a less mutually dependent nature do exist.

An alternative, interesting and persuasive view of supply chain management is given by Cox and Ireland (2002) in terms of the appropriateness of certain relationships according to the firm's power position within the supply chain. Here, there is only scope to severely précis their argument. They believe that there is 'a low level of trust in the buyer–supplier exchange relationship because of the frequent asymmetry of the buyer and seller power attribute'. By understanding this relationship, it is possible for buyers and suppliers to know the most appropriate relationship management approach available to them. Buyers with power (e.g. those which build frequently) have the ability to develop their supplies competence and transform the supply offering directly. The corollary is supplier power when there is only one or a limited number of suppliers, and power which, at least theoretically, vests with suppliers. Rather than reflecting the search for alliances that are intended to build trust and co-operation in a non-adversarial relationship, the power scenario reflects the opportunistic phenomena of transaction cost economics that the subservient partner will seek to protect against. The power scenario is, perhaps, a more realistic image of many construction industry relationships.

An initiative to put integrated supply chain management into practice is the 'building down barriers' programme instigated in 1997 by the UK government's Department of the Environment, Transport and the Regions combined with the Ministry of Defence and two major contractors. One element of the programme was to adopt a project structure based on 'work clusters' as reported on by Nicolini *et al.* (2001). They state that a cluster is 'in many respects, a "design and construct" mini-project that takes place within the larger framework of the project. It is a place where a limited number of designers, suppliers of materials or components engage in intensive collaboration to design and deliver a significant, recognisable element to the overall building, working to reduce costs, improve value and minimise waste. Clusters operate as semi-independent parts of the project under overall co-ordination of the contractor or project manager, replicating the logic of one-point responsibility down the supply chain'.

The benefits were seen to be 'the potential for the clustering system to provide a structure within which key suppliers can take initiatives to optimise their portion of the design and their construction processes' and 'permitted specialist subcontractors to take a far greater role in design development than would normally be the case in a design and build contract'. A further comment was that 'Involving the clusters in design development has also, on the whole, been successful as a route for the contractor to negotiate firm and realistic prices with key suppliers while obtaining a fair profit level'. Problems were associated with whether design responsibility had really been passed to the clusters or remained with the design consultants, reflecting established patterns of deference to design consultants. A major issue in this respect was the position regarding professional indemnity insurance which was held by the consultants.

Trust between construction organisations generally

Trust has emerged as an issue for serious study in construction, mainly as a result of the development of partnering and supply chain management; and this is reflected in the earlier discourse on these topics. Put simply, the argument there was that such relational initiatives were designed to reduce transaction costs by reliance on trust engendered by a change in the culture of project participants rather than by protecting against opportunism through contracts.

The focus was on relations between the client and contractors, subcontractors and suppliers and also between contractors, subcontractors and suppliers. On traditionally arranged construction projects and their derivatives, it is assumed that these parties will behave opportunistically and, to protect against this, their relationships are closely prescribed by detailed contracts leading frequently to adversarial relations. It is recognised that contracts and trust can and do substitute for one another and that contracts may act as a double-edged swords. On the one hand, they prescribe how organisations should co-operate to reduce the risk of exploitation; on the other hand, they can make it difficult for trust to develop (Murnighan *et al.* 2004).

But what of the relationships between clients and other members of the project team, on both traditionally and relationally arranged projects? To what extent do trust and opportunistic behaviour determine their relationships? Trust is seen as an 'ambiguous and complex phenomenon' (Kadfors 2004), and many definitions have emerged; but for relations in construction, the one used earlier remains appropriate:

> 'The willingness of a party to be vulnerable to the actions of another party based on the expectation that the other will perform a particular action important to the trustor, irrespective of the ability to monitor or control that other party.' (Johnson-George & Swap 1982)

The perspectives taken on trust are vast, but here the types of trust considered are those defined by Kadfors (2004) as rational choice (based on transaction cost theory and hence opportunism), relational trust (between individuals, reflecting the vulnerability in the above definition which recognises that there are psychological and social risks as well as financial risks) and institution-based trust (the role of institutions in creating the conditions necessary for trust to arise, for example legal systems, societal norms and, in construction particularly, professional institutions).

The relationships between the client and project manager, architect, engineers, quantity surveyor and other professionals; between the project manager and these professional contributors; and between the professional themselves, all require elements of trust to make them effective. In the past, when professionals were appointed and paid on the basis of a scale of fees which specified their duties (often not very precisely), there existed a high level of trust based on professional standards decreed by professional bodies (institutional trust) and professional ethics. Subsequently, scale fees were abolished by the Monopolies Commission and more commercial approaches prevailed as clients required bids for commissions and entered into more tightly drawn contracts with the professional contributors. These events increased transaction costs (contract preparation and negotiation to guard against opportunism), reducing the need to rely on trust. Nevertheless, the relationship between clients and the professional contributors continues to rely on a significant level of trust and professional ethics, as the ability to specify all contingencies in contracts is limited and the contributors have a vested interest in maintaining relationships with the prospect of further commissions. This is particularly so as in many cases the lowest price may not have to be accepted and confidence (trust) in a specific bidder may outweigh the difference in bids. The relationship between the contributors themselves is, of course, not based on direct contracts but relies on relational trust, with institutional trust having an influence on how they respond to each other. Similarly, and particularly importantly, the relationship between the project manager and the contributors is based on relational trust and requires the project manager and the contributors to nurture confidence in each other. This is particularly challenging for the temporary organisations required for construction, for which collaboration on previous projects is extremely valuable.

The type of situation described above prevails when the contributors are not in-house to the client organisation but are drawn from outside firms. For projects where the client has an in-house capability, mutual trust between all parties is relational. Wong *et al.* (2000) investigated the factors that influence the level of trust between contributors to projects within a public sector infrastructure organisation; intra-organisational trust. The factors (or antecedents) generating trust which were tested were drawn from Shaw (1997). They were: achieving results (competence), acting with integrity, and demonstrating concern, and each was found to be causally related to trust. Wang *et al.* make the important point that 'sustaining an appropriate level of trust requires the judicious balancing of the three antecedents, even when they come into contact with each other. This places particular demands on the project manager'. They go on to say that

> 'the balancing act requires enlightened management, compatible organisational structures and processes and appropriate organisation culture. This contrasts dramatically with the increasing demands for competition in an ever-changing global economy, rendering commitments increasingly short-lived as conditions call for rapid changes in an organisation's strategy and policy. In this case, the emphasis on results may dominate that of concern for the well-being of the participants. Similar situations may arise in times of crisis, process re-engineering or even relatively simple changes in organisation structure'.

An aspect which seems to be downplayed in the literature on trust is the actual personal characteristics of each individual person who is called upon to trust or not to trust in a relationship. Individual people do have different propensities to trust: some are trusting, some not so trusting and some distrustful. The corollary of this is that some are trustworthy and others are untrustworthy, with a range between the extremes. In the close professional co-operation which is necessary for successful construction projects, the perception of such characteristics in others can have a fundamental effect on the quality of collaboration and hence on the success of project outcomes. McDermott *et al.* (2005) recognised this factor and highlighted the importance of key relationships to the success of a project.

Chapter 7
A Model of the Construction Process

Introduction

Now is the time to draw together the themes that have been running through the previous chapters into a model of the construction process. This approach formalises the ideas that underpin the way in which construction project organisations should be structured and provides an approach to analysing and designing project management structures. Although the approach may at first appear theoretical, it does provide the basis of a practical, analytical tool for examining the effectiveness of the project management process, as described in Chapter 11.

The construction process has few fundamental characteristics common to all projects. This is not unexpected in view of the diversity of construction projects and their clients. That being the case, it is necessary to identify those aspects that can be generalised so that they may then be interpreted for each individual project. The application of the model will then identify the structure of the process in such a way that it is possible to analyse how it operates in practice.

Such a model may be employed as a tool for learning from experience in a more rigorous way than has been the case in the past, by using it to analyse completed projects. But, more importantly, it can also be used for designing organisation structures with the aim of providing the structure which should give the best chance of a successful project outcome as far as the organisational aspects are concerned. The tasks and roles of project management can then be identified on the basis of the organisational structure designed for each specific project. Project success is, of course, dependent upon much more than solely organisational issues, such as behavioural, political and other forces acting upon the project, but if the organisation structure is as well designed as possible, at least the project is off to a good start.

Common characteristics

A prerequisite of the model is an outline of the process of providing a project devoid of artificial organisation boundaries such as those created by conventional and other predetermined approaches to project organisation. Such a model would

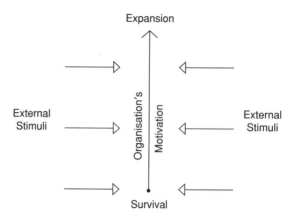

Fig. 7.1   Organisation's response to external influences.

identify the major forces that influence the process and the fundamental structure that results.

The process has a start point (which may be difficult to identify specifically in practice). It also has a finish point, which is taken as the completion of a project. The process of identifying and providing a project consists of those events that join these two points. Potential start points are activated by organisations which *may* become clients of the construction industry if the process identifies that a construction project is required to meet the objectives of the potential client. The term 'client' is used to refer to a sponsor of construction work who can generate the finance, information and authority necessary to embark upon the process.

Construction projects start as a result of the influence of environmental stimuli upon prospective client organisations which create the motivation and need or opportunity to construct to reach objectives. Such stimuli may be economic, technological, sociological, etc. and usually consist of combinations of different classes of forces. The basic response of an organisation to environmental stimuli is the result of its need to survive; above this level the organisation responds in order to expand as the result of its motivation (see Fig. 7.1). Survival is the basic goal which requires the organisation to maintain its position relative to those of its competitors for which it must continue to obtain a return acceptable to its environment in terms of its role (e.g. profit, service, acceptability). This is more easily conceived for commercial organisations, but is also true for public authorities.

Public authorities have to adapt to survive and also strive to expand by increasing the quality of their services. If they are perceived not to be doing so other means of providing their services may be found. Privatisation of previously public services in many western economies is evidence of this although many such changes have not necessarily been as a result of economic forces as proposed by the transaction cost approach but sometimes as a result of political pressures.

Expansion is a response to environmental forces by the organisation to take advantage of events in its environment. The degree to which the organisation takes such opportunities is determined by its motivation, which is, in turn, influenced by incentives provided by the environment, e.g. taxation, status, satisfaction. The start point of the process is, therefore, the recognition by the potential client of the need or opportunity to achieve a particular objective for their organisation. The options available to achieve the objective *may* include the acquisition of real property (defined here as encompassing both building and civil engineering outputs) which, in turn, *may* require the construction of a new project, but at this early stage this will not have been established.

At the initial activation of a start point, the plane within which a finish point is feasible will be very wide and will encompass all those alternatives that allow the organisation to achieve its objective. The alternatives available will vary, depending upon the nature of the organisation's role. For instance, there will be basic differences between the choices available to commercial and public authorities. However, *it is possible*, for every category, that one of the alternatives will require the acquisition of real property.

This outline of the process is now illustrated further using as an example the identification and provision of a building for a commercial organisation. The concepts are more readily understood in terms of commercial criteria, although the same process should be followed by any potential client and for processes which lead to either a building or civil engineering project as the outcome.

After starting the process, the initial decision of relevance to the construction process is whether or not real property is required. This may be called the *project conception process*, as illustrated in Fig. 7.2. If a decision is reached that does not require the construction of a project, then the organisation which was a potential client of the construction industry will not become a client. During this phase, environmental influences are transmitted to the potential client through the importation of information, energy and materials from the environment. The meaning of information and material is self-evident, although it should be pointed out that material encompasses any material whatsoever. Energy similarly means any type of energy, but in this context people are a particularly important source of both physical and mental energy. Such influences can be broadly classified as political, legal, economic, institutional, sociological and technical. The action of these influences will determine the initial decision. The project conception process will entail the consideration of each alternative within the environmental context and a decision will be made on the basis of the influence of the external factors. For example, economic conditions may make a process change appropriate, but it may then be discovered that rapid technological development makes change inappropriate at this time, by which time economic conditions may have made the take-over of another firm more appropriate. This process is one of the client organisation adapting to environmental influences until an initial feasible decision is reached and normally takes place within the client organisation.

In developing the model, it is assumed that the preferred outcome of the project conception stage requires the acquisition of real property to contribute to the satisfaction of the potential client's objective. At this stage, acquisition of real

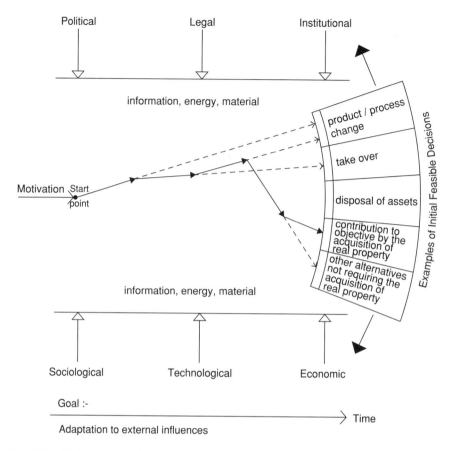

Fig. 7.2   Project conception process.

property includes existing or new property or improvement or modification of property already owned.

The assumed preferred decision of the project conception process, which requires the acquisition of real property, contains a number of alternatives which can be considered as an intermediate feasible decision point. The process of arriving at one of these alternatives in making further progress towards the finish point may be called the *project inception process*, and is illustrated in Fig. 7.3. The intermediate feasible decision actually made is again determined by the ability of the alternative chosen to contribute to the achievement of the objective of what is now a client of the construction process. The environmental influences acting upon the process of reaching an intermediate decision are the same as those given before, but may exert different influences during this process. The project inception process will receive information, energy and material from the environment and will transform them in its task of identifying the appropriate intermediate decision. Interacting with these influences in arriving at a decision

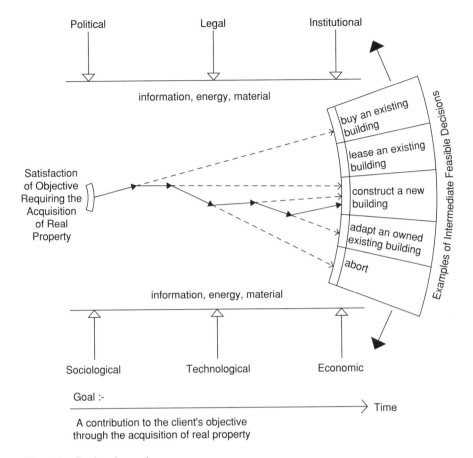

Fig. 7.3   Project inception process.

will be the commercial activity of the client, which will itself be influenced by the external factors. For example, the environment factors affecting the decision will be the state of the property market regarding the availability of existing premises and rent levels, the cost of new building work, site availability, the rate of technological change which determines whether a short-term lease is better than a freehold building, sustainability issues, etc.

In developing the model further, it is assumed that the preferred outcome of the project inception process is the construction of a new building. The performance of the building that is actually constructed will lie within a finish spectrum ranging from total satisfaction with performance requirement to total dissatisfaction. The process of arriving at the finished building is called the *project realisation process*, as illustrated in Fig. 7.4. This will again be determined by the environmental influences acting on the process. These are classified as before, and again provide information, energy and material for the process. For example, the environment provides the skills, both professional and constructional, which

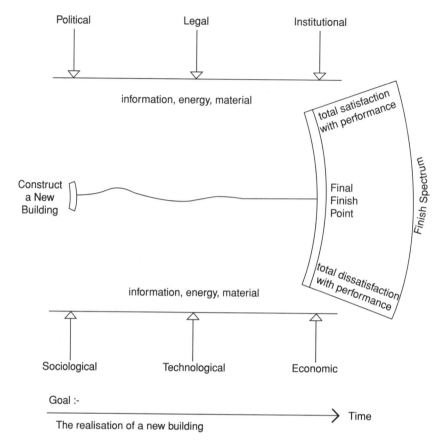

Fig. 7.4   Project realisation process.

are available to the project. It also determines the availability of materials, and even the weather, which might affect the completion time for the building. The client's environment will affect the certainty of what is required of the building. If it creates uncertainty, this might generate changes to the design and construction which can affect the cost and completion time for the project. As was the case with the project conception and inception processes, the external influences act in two ways: directly upon the process and indirectly through their influence upon the commercial activity of the client. The project realisation process transforms these inputs into the output of the process, which is the finished building. The effectiveness of the transformation process will determine the quality of the outcome actually achieved.

   A further example of the effect of external influences during this process could be that economic and/or institutional forces determine that construction work is awarded on the basis of a competitive tender. Such a decision would divide this process into two sequential sub-processes, design and construction, but only if appropriate external influences are present. Such an assumption would be unfounded at this stage of development of the outline of the process.

To summarise, there are only three subsystems that have universal application to all construction projects: the project conception, project inception and project realisation subsystems. These subsystems generate two primary decision points, one which contains the potential decision that real property is required and, if such a decision is taken, the second division as to the nature of the real property to be provided. Both these decisions will be taken by the client organisation.

The start point represents the beginning of the project management process and it can be seen that this will, in most cases, be contained and managed within the client organisation during the project conception subsystem. Ideally, the client organisation should, during this process, receive advice from the project team, but in reality this rarely happens. If it were to occur, then the members of the construction team involved would be part of the project management process but would not be leading it, as at this stage a business decision rather than a construction decision is required. However, it would certainly be advantageous for clients to have an input of advice from certain members of the construction team at this stage to enable them to take a fully informed decision.

The project inception subsystem demands a significant input from the project team and the process will require a property or construction orientated management system working in conjunction with the client to identify the most appropriate solution to the client's needs. However, project management does not usually take place in this form in practice. It tends to occur implicitly, solely within the client organisation, and project teams are often faced with a *fait accompli* by the client. This may be appropriate when the client has in-house expertise but to the disadvantage of the client when this is not the case. Project teams can, therefore, often do little other than proceed on the basis of the client's preconceived idea of the best solution to the problem. It would be advantageous to both the client, in terms of the utility of the completed building, and the project team, through their increased effectiveness, if clients were to involve members of the project team in this process.

The project realisation subsystem is the process that is most readily perceived by clients and project teams as that in which project management takes place, although even in this process it is sometimes construed as being concerned only with the construction phase rather than with the whole process, including design. Even though project management may be led by someone from the project team during this process, the management of the process will benefit significantly from the involvement of the client.

As there are only three subsystems that are universally applicable to construction projects, it is necessary to identify the factors that determine the subsystems within each of these major subsystems which are identified by the two primary decisions discussed previously. As most projects differ, the subsystems required to achieve them will also differ as a result of the task being undertaken and the environment within which it is carried out. Therefore further subsystems cannot be defined explicitly, but what is possible is to identify the factors that create them, the nature of their relationships and their need for integration.

Subsystems

The primary decision points differentiate the major subsystems and define the boundaries between them. Similarly, other decision points will determine the boundaries between other subsystems. The construction process is characterised by a series of decision points. These act as 'pinch points' through which the project team must pass if progress is to be made. If the project cannot pass a pinch point, then it will be aborted. Decision points are arranged in a hierarchy of which the primary decision points are at the highest level. Below them occur key decision points and operational decision points, the distinction between which Bennett (1991) has confirmed as being especially important in providing the basis for integration.

Key decision points

Further discontinuity in the system is created by key decision points. Key decisions are those that the clients will make for themselves. They are determined by environmental influences acting upon the client organisation and are often manifest in the client's internal procedures for expenditure and similar approvals. They can range from, for example, approval of design and budget proposals and decisions to delay the project, to decisions to change the nature of the project. Such decisions imply a degree of irrevocability, as to revoke such decisions would entail the client in a loss of resources. Bennett (1991) confirms that experienced clients often ask for a formal report at key decision points.

The process of providing a project is characterised by discontinuity created by the need for decisions. As they reflect the flow of the process they are fundamental to the organisation structure of the project. The integrating mechanism provided between the client and the members of the project team is, therefore, highly significant for the success of the project. This mechanism should aim to anticipate key decision points and organise so that decisions can be made on the basis of the contribution of the members of the project team, who have a part to play in providing information and advice on which such decisions should be based. The nature of the client's organisation can have a fundamental influence on the effectiveness of this process. If the client is represented by a committee or board of directors, they will have to decide how they themselves are to take key decisions or whether they will delegate this function to some member of their own organisation and, if so, to what extent they will delegate. Similarly, they will have to decide upon the extent of delegated authority given to the project team as this will determine which are the key decisions. Decisions taken by the project team are classified as *operational decisions*.

The greater the number of decisions classified as operational, the greater will be the flexibility available to the project team and the more control they will have over the decision-making process. There is likely to be less uncertainty and delay and more integration, for example, in those cases where a client's representative has sufficient delegated authority to make many key decisions, or in cases where the project team deals directly with, say, a managing director acting with full authority. This is likely to be the situation in the former case only where the client has great confidence in the representative and, in the latter case, for the smaller private company. Very real difficulties can be created for the larger private companies and public authorities if an appropriate integrating mechanism is not or cannot be designed. A common problem is that of key decisions taking longer than anticipated, with a resultant delay to the project, which can often have a corresponding knock-on effect for later project activities. A further common problem is that if the client's organisation is unresponsive to environmental forces and the needs of the construction process, key decision points may be inappropriately identified in terms of the nature of the decision and its timing. An important task of the leader of the project team is, therefore, to endeavour to make clear to the client the team's needs in terms of the timing and quality of key decisions. Yet, at the same time, the leader must recognise that the project will place heavy demands upon the client's organisation, which will have to continue to carry out its main function as well as being involved in the design and construction of the project.

Key decisions lie beneath the primary decisions in the decision hierarchy and, therefore, contribute to them. Until a primary decision point is reached and the decision is taken, key decisions will be cumulative. They therefore provide major feedback opportunities for both the client and the project team. Each key decision should be checked against the overall objective of the project and the primary and key decisions already taken, to ensure that the project is remaining on course. If it is not, then it will either have to be returned to course or the original objectives or previous decisions will have to be reassessed to establish whether the deviation is beneficial and feasible. Whereas the primary decisions create the boundaries between the major systems of the process, the key decisions represent the boundaries between the main subsystems which constitute the major systems and provide feedback opportunities as shown in Fig. 7.5.

Operational decision points

In bringing forward propositions upon which key decisions will be based, the members of the project team will themselves have to make decisions based upon their professional and technical competence. These will not affect the policy of the client's organisation: such decisions will be primary and key decisions, and will be taken by the client. Therefore the decisions taken by the project team in making progress towards a key decision are described as operational. The range of propositions that may be presented to a client, and the operational decisions implicit in defining the propositions, will be the responsibility of the leader of the project team. The leader will also have the responsibility for integrating the project team to ensure that all relevant advice has been given before arriving at an operational decision. The leader must also make certain that the full range of propositions appropriate to the client's objectives have been formulated and that they are presented objectively at key decision points. Unfortunately in practice, on many projects clients are not presented with alternative propositions and in

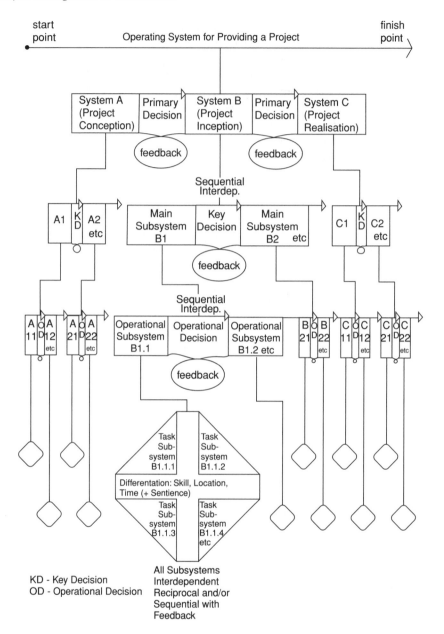

Fig. 7.5   The operating system.

such cases operational decisions are made solely to move the process towards the next key decision point at which the client is asked for a 'yes' or 'no' decision to the single proposition.

As with key decision points, operational decision points will also represent 'pinch points' through which the project must pass if progress is to be made.

Although they do not have the same degree of irrevocability and associated loss of resources as key decisions, nevertheless operational decisions which are later changed are likely to cause delay and some loss of resources. Examples of operational decisions include the details of project programmes, the use of bills of quantities for tender documentation, and alternative technical design proposals. These decisions are therefore mainly concerned with professional decisions and with the implementation of procedural aspects of projects, and move the project incrementally towards a key decision. Their position in the system will be determined by previously taken key decisions, but there will be more opportunity for the project team to design the structure of these decision points than is the case with key decision points.

Operational decision points present secondary feedback opportunities. Each time an operational decision is being considered, it should be checked against the previous operational decisions within the subsystem and against the last key decision to ensure that it is compatible with what has been decided previously. If it is incompatible, opportunity to change the decision exists or, alternatively, to amend a previous operational or key decision, although this is unlikely to be the outcome at an operational decision point.

Each subsystem created by key decisions will, therefore, consist of a number of operational subsystems, as illustrated in Fig. 7.5. The number of subsystems created by key decisions and operational decisions cannot be prescribed for all projects as they will vary considerably in both number and nature depending upon the type of project required and the environmental conditions in which it is to be achieved. However, the model provides a basis for identifying them for individual projects.

Task subsystems

In arriving at an operational decision, each operational subsystem will consist of at least one task subsystem. The task subsystem level is where the contributors to the project work together to bring forward the propositions upon which operational decisions are based. At operational and key decision level it will be the job of the manager of the process from both the client's organisation and the project team to bring together the decisions in a manner compatible with the client's objectives.

The following illustrates the task subsystem level for an operational subsystem concerned with 'identifying a site' at the same time as which the outline requirements of the client are being finalised and feasibility studies also have to be carried out. This would require three task subsystems to be operating with reciprocal interdependency. Various people would be carrying out the three tasks and taking advice from a range of different people. For instance, the managing director of the client organisation could be finalising outline requirements with advice from departmental managers and, perhaps, some representatives of the project team; a commercial property agent could be trying to identify suitable sites with advice, say, from the managing director, architect and quantity surveyor; and the feasibility studies would be carried out by the quantity surveyor with advice from the property agent, architect, structural engineer and managing director. The result of this interaction should be propositions which could result in an operational decision to enter negotiations for sites prior to a key decision to purchase a particular site. The problem of managing this type of process is one of integrating such a diverse range of interdependent contributors. In modelling this element in generalised terms, therefore, it is necessary to identify the features that generate the differentiation requiring integration.

The determinants of differentiation have been expressed by Miller (1959) as technology, territory and time and have general relevance to the construction process. The idea of differentiation, as described in the last chapter, provides the tools for analysing different situations that give rise to various levels of differentiation and hence to the need for matching levels of integration. An understanding of the degree of differentiation present in a system significantly aids the person managing the system to provide the appropriate level and mechanisms of integration.

Separation of people working together on a project is affected by the skills (*technology*) they bring to bear on the project. People from different skill backgrounds (e.g. architect, engineer, quantity surveyor, builder) work on problems in different ways, which stem from their experience, and they often cannot see the other contributors' points of view. Such conflicts need reconciling by the manager of the system in terms of the client's objectives. Similarly, separation on the basis of location (*territory*) creates boundaries between contributors. Territory refers to the geographical distance between groups of people working on the project and this obviously affects communication to a very large extent. Even though IT has made communication easier, particularly of technical matters, the most intransigent problems, of which there are usually many, require iteration. How much easier it is to resolve such problems on a project through face-to-face discussion. The advantages to be gained through the contributors working together in the same office (although from different firms), particularly during design, are likely to be significant. A further differentiator is *time*. Although this was visualised by Miller in relation to shift working, it is relevant to the contributors to construction projects in terms of the sequence of activities to be performed where particular contributors cannot perform their activity until another has completed their own.

Overlying and reinforcing differentiation on the basis of the above is sentience, also referred to in the last chapter. It is a particularly strong force in the construction industry and gives rise to strong allegiances to a profession or a firm or to both.

Figure 7.5 illustrates that each operational subsystem will consist of task subsystems and the people working within them will be differentiated on the basis of skill, location and time, reinforced by sentience. The number and nature of the task subsystems and the people working within them have to be defined for each individual project as they will vary significantly.

It is perhaps necessary to comment on the relationship of the skill (technology) of project teams and the technological complexity of the project. The skills (technology) needed to design and construct a project are, of course, determined by the technology of the project, that is the skills required for a simple primary school are quite different from those required for the Channel Tunnel. Hence the more technologically complex a project, the greater the differentiation and the need for higher levels of co-ordination (Scott 1992). Bennett (1991) also points out that only when managers and designers have worked out the nature and character of innovative projects can further elements of the project team be added. Therefore, for projects which have standard solutions (i.e. technologically simple), a complete project organisation can be established early but when the technological solution requires much development full teams are established later. Bennett's views on project size are consistent with those on technology. Thus views on the importance of technology and size are consistent with the decision framework described above, as the scale of a project and the difficulty of the technical definition will be reflected in the decision structure. Technology and size give insights to the reasons for certain decisions, and contribute to the nature of the key and operational decisions.

The operating system and the managing system

What has been defined so far in the model is the operating system – that system of activity through which the project is achieved. Figures 7.2 to 7.5 model, in abstract terms, the generality of the construction process. The operating system is managed by a managing system, which carries out the decision-making, maintenance and regulatory activities that keep the operating system going. It is differentiated on the basis of skill from the operating system. The skill of the managing system is management and those of the operating system are professional and technical. The managing system referred to here is that which acts on behalf of the client. It is concerned with the totality of the process of providing the project, which includes that part of the client's organisation relevant to the process. Each system and subsystem into which the process is differentiated may have its own managing system, but such systems will not be managing the total system for the client. The concept of operating systems and managing systems has also been supported by empirical work concerned with construction on site (Shirazi *et al.* 1996).

The actual form which the managing system takes in practice will vary considerably. It may be contained solely within the client's organisation, where the client has this capability; in other cases it may consist of a client's representative and a 'consultant' project manager. In the conventional arrangement of contributors it would be the architect acting in a dual role of manager and designer. Alternatively, it may be undertaken by a variety of people at different stages of the project.

The managing system controls the boundaries between the systems and subsystems and integrates their output to ensure that the primary and key decisions made at these boundaries are compatible with the client's requirements. The managing system should ensure that boundaries are appropriately drawn in relation to the process, that facilities for appropriate feedback are available and are used, and that the correct decisions are taken. To achieve this, the managing system also seeks to control the boundaries between the process and its environment, and between the process and the client and its environment.

In order to support this role, the managing system needs to monitor the performance of the systems and subsystems. Such intra-system regulatory activities are intended to ensure that the manner by which systems and subsystems arrive at the propositions upon which decisions are based is appropriate. This entails the design and use of feedback mechanisms and requires the managing system to integrate the subsystems and to ensure that appropriate techniques are used. Although monitoring activities will also be carried out by the systems' and subsystems' own managers, nevertheless the managing system of the total process acting for the client will need to convince itself that the operating system is using appropriate methods.

The managing system also needs to ensure that the resources that produce the output of the systems and subsystems (for example, and in particular, people) are procured and replenished. These activities aim to ensure that the operating system has the capacity quantitatively to perform its tasks. Such activities will also be carried out by the managers of the systems and subsystems, but the managing system of the total process acting for the client will again need to determine the fact that the operating system has the capacity to perform its tasks.

The functions of the managing system

The range of activities required to be carried out by the operating system varies between projects depending on the nature of the project, its environmental context and the consequent arrangement of decision points. The detailed work of the managing system will also vary but, nevertheless, its functions can be conceived and generalised irrespective of the nature and organisation structure of the project. These functions are identified and discussed first before translating them into the detailed functions of project management and the range of skills required by a project manager.

Management functions are exercised over the people carrying out each of the tasks making up the project process but more importantly they are concerned with managing the interrelationships of the tasks (or, put another way, with managing the 'space' between people and between tasks), and with managing the relationship of the project to its environment.

Approval and recommendation

Perhaps the most important relationship within the managing system is the connection between the power of approval and the right to make recommendations. The power of approval is, of course, exercised at the decision points in making decisions. The right to make recommendations refers to the authority to make a specific recommendation or to present the alternatives upon which a decision will be based. A person with this role is in a very influential position to persuade the person with approval powers to make a particular decision and so select the alternative that the person with recommendation powers wants.

The managing system normally consists of at least two components, one representing the client and the other the person managing the project for the client. The former will normally be a member of the client's organisation and the latter a member of the project team employed for the specific project. The pattern of approval and recommendation powers between them will depend upon the role the client decides to take and the structure of their organisation. For example, on three projects analysed using the model (Walker & Hughes 1984, 1986, 1987a), and which remain valid today, clients reserved for themselves approval of the output of most tasks up to commencement of construction, with the exception of a small number of tasks that did not involve choices between alternatives, e.g. preparing contract documentation. The level at which the approval powers were vested in the client organisation's hierarchy depended upon the structure of that organisation. For example, for one project the early decisions were approved by local directors of the parent company until the basic parameters had been established. Then approval powers passed to the client's in-house project engineer. Subsequently, the directors were only involved in approvals at a limited number of decision points. Then, for the construction phase, approval powers passed to the manager of the project team (who was titled project manager and was employed by the engineering consultants). These powers did not, however, include responsibility for approval of further project instructions and of documentation, including drawings, produced by the design team, which the client's project engineer approved. The project manager had the role of recommending courses of action for the client's approval which included presenting and advising upon the choices available. It was this activity from which the project manager's authority on the project was derived. The project manager approved proposals of the contributors, but the final approval to proceed remained with the client.

For two other projects the client's organisation structure was simpler, with the group chairman and managing director respectively representing the client throughout the project; they personally retained approval powers for all decisions during design but not during construction. Again, the manager of the project team recommended actions to the client except during construction, when the manager had approval powers and also had responsibility for approving a small number of routine tasks during design. For these projects additional design information during construction was not subject to approval by the client. The management of the project again drew authority from the power of recommendation.

These arrangements illustrate clients' wishes to be closely associated with their projects and an unwillingness to delegate approval powers. The opportunity for public sector clients to adopt the same position is much more difficult, particularly if the client takes the form of a committee. In this situation the 'client committee' is likely to make one of its officers responsible for managing the project as one of the components of the managing system. This can produce a complex situation regarding approval powers and may result in them being split three ways, part with the 'committee', part with the manager from the client organisation, and part with the manager from the project team.

These issues demonstrate the importance of integration between the client's organisation and the process of construction. It is clear that the client will determine the approval pattern within a project and hence define the authority of its managers of the project. The degree to which this can be formalised will depend upon the will of the client as influenced by the managers. Although it would be ideal to have approval and recommendation powers written down in an explicit form, in reality it is difficult to persuade clients to do this adequately. Nevertheless, it is an ideal for which managers should continue to strive. However, even if it is achieved, there will still remain the problem of interpretation.

The key to these relationships is understanding and communication between the parties involved, which cannot be fully dependent upon formality but which will draw strength from the informal integration of the people concerned through discussion of the issues surrounding the project. In all probability the relationship between approval and recommendation powers will evolve during the project as trust and understanding develop between the participants. Nevertheless, efforts should be made by the client and managers of the project to establish guidelines for these relationships as soon as possible in the project's development.

Boundary control, monitoring and maintenance

Boundary control, monitoring and maintenance are the systems terms which represent the basic project management activities carried out by managing systems. These activities are normally carried out by the component of the managing system drawn from the project teams, although if clients' organisations have the capability, they could be carried out by their people.

The objective of boundary control is to ensure functional compatibility of contributors' work within and between tasks, to relate the system to its environment and to control the system's direction towards the required outcome. It is fundamental to the achievement of the level of integration and control demanded by a project and to a satisfactory project outcome. Boundary control derives its name from its role in managing the boundaries between the various subsystems of the project.

This activity is normally accompanied by the complementary activities of monitoring and maintenance. Monitoring is intra-task regulation to check and control prior to output that a task is proceeding in a manner which will achieve its purpose. Maintenance ensures that a task has the capability to achieve that purpose.

Boundary control involves setting up formal control mechanisms using the feedback loops defined by the key and operational decision points identified when designing the organisation structure, and establishing the supporting information system. It ensures that information flows as intended and that feedback mechanisms are activated. In addition, boundary control should ensure that the reciprocal and sequential interdependencies identified in designing the organisation structure are made to work in the manner intended.

Sequential interdependencies can be integrated by ensuring proper information flow in accordance with the information system, but reciprocal interdependencies need to be integrated using mechanisms that ensure that contributors meet in the correct combinations and at the right times. Such mechanisms would normally include action-minuted meetings and exploratory and less formal meetings in the critical early stages of the project. Where contributors are from different firms, it could, and in many cases should, extend to bringing the people together in one place to work on the project rather than their relying on correspondence and electronic communication. These activities include ensuring that the client is integrated in the appropriate manner at the various stages, and keeping in close contact with the client to identify any changes in environment that may affect the client's requirements. Whereas boundary control relates the parts of the system to each other in the way described earlier, monitoring seeks to ensure that the individuals or groups undertaking a specific task respond to the demands to integrate and also that techniques and procedures appropriate to the specific task are being used.

Maintenance involves keeping in close touch with each contributor and ensuring that each is equipped to carry out the task required. It requires regular formal reviews of the quality and quantity of resources dedicated to the project, particularly in relation to the number and level of skill of the people employed on the project.

Boundary control, monitoring and maintenance are managing system activities and, in accordance with the proposition that the managing and operating systems should be differentiated (on the basis of the skills needed), they should be vested in someone who is not also undertaking operating system activities on the project.

These ideas of what the managing system should be doing sit uncomfortably upon the way in which the construction industry and its professions have evolved. They demand a much greater involvement, some may say interference, by the managing system for the project (e.g. by the project manager) in the activities and tasks carried out by the contributors. The evolution of the industry and its professions has resulted in the creation of a large number of independent firms, but they are interdependent when working together on a project. These ideas, therefore, suggest that control over such firms, both in terms of their relationships with each other and, more significantly, in terms of the activities within a firm, are a legitimate activity of the project management process.

If the managing system is to control the construction process satisfactorily, it requires the authority to carry out boundary control, monitoring and maintenance activities in connection with the activities of all the contributors. The contributors would have to be prepared to accept such authority while still remaining responsible for their individual input, whereas the managing system should be responsible for overall project control. This web of formal relationships is one of the most difficult aspects of structuring a project organisation and one which is not readily faced by clients and project teams. The managing system needs to ensure that it is properly worked out, documented and understood by the participants at an early stage in the process.

Notwithstanding the need for structural relationships to be defined, in reality the effectiveness of the project team will depend ultimately upon the informal relationships generated by the managing system. They will be the product of the way in which it deals with the project team in carrying out its boundary control, monitoring and maintenance activities. Its objectives should be to weld the contributors into a true team and to ensure that they recognise that satisfaction of the client's objectives should be synonymous with satisfaction of their own.

General and direct oversight

There are two further classes of supervision of relevance to the project management level of control, general and direct oversight. Although these are not project management activities as such, the concepts are directly relevant to the effectiveness of the project management process. This does not mean that other managing activities at lower levels in the project hierarchy do not have implications for project management, but they would be the responsibility of the contributing firms and be overseen by the managing system through monitoring and maintenance activities.

General oversight provides policy guidance for the project and *direct oversight* is concerned with directly supervising specific skills used on the project. The manner by which these activities are distributed among the project team depends upon the structure of the firms that contribute to the project organisation. In the case of general oversight (policy guidance), this will often be exercised by the client in the conception process of the project until its broad outlines have been approved. The actual person appointed to undertake this activity depends upon the structure of the client's organisation. If the client's component of the project managing system is provided by a person in authority in the client's organisation, e.g. the managing director, then that person will normally be providing general oversight as well as being part of the managing system. In the early stages of the project, before the project team is involved, that person will be guiding other members of the organisation as the ideas for the development of the project are generated. In such a situation the appointed person may then also become part of the managing system when the project team becomes involved. On the other hand, if the client is represented by a committee, it will provide general oversight. If, as is probable, it delegates to a member of its organisation responsibility for managing the project in conjunction with the project team, the committee will continue to give guidance on policy matters.

A range of possible arrangements are available, and the one selected will be a function of the client's organisation structure. What is important is to understand who is exercising which function.

As the project progresses, the policy guidance will probably pass from the higher management levels of the client's organisation to lower levels. For example, it may pass from the board of directors to the client's 'in-house' project engineer who will become responsible for general oversight as well as being a component in the managing system.

Subsequently, general oversight may pass to the project team when the detailed work on the project commences. The person who then exercises it depends upon the structure of the firm providing the management of the project. For example, if this function is provided by a firm of project managers and the actual project manager is not a partner in the firm, then a partner may provide policy guidance on behalf of the client. The degree to which general oversight on behalf of the client is exercised by someone in this position will depend upon the degree to which the client is prepared to delegate at this stage. If the project manager is also a partner in the firm, he or she will probably exercise both general oversight and project management activities.

A danger at the stage of detailed implementation of the project arises if the client leaves the project team to 'get on with it'. If the managing system is not properly structured and consequently the contributors pursue their work relatively independently, policy guidance may not be provided and the managing system will not be structured in such a way that the lack of guidance is recognised.

The manner by which direct oversight is provided again depends upon the structure of the contributing organisations. Direct oversight is the highest level of supervision exercised over the individual skills used on the project, for example by a partner of one of the contributing professional firms over the activities of people working on the project. Similarly, the relationships between direct supervision and project management depend upon this structure.

For example, where all professional skills and the components of the managing system from the project team are provided by a multidisciplinary firm, direct oversight will be provided by the departmental managers of the firm for each particular skill. Such managers may also be partners of the firm but would not also be acting as project managers if the managing and operating systems are kept separate. Some other member of the firm would be acting as project manager and the relationship between project management and direct oversight in the matrix should be established within the firm. If the contributors are from separate professional firms, their partners will carry out direct oversight, and the relationship with project management, if also provided by a separate firm, will need to be established and could be potentially more difficult to achieve.

Many other arrangements could exist but, again, the important thing is to recognise where responsibilities lie. In many cases, professionally qualified members of contributing firms do not require direct supervision, but this depends upon their status and the policy of the firm by which they are employed.

Because of the continuing use of competitive tenders for construction work and the consequent standard conditions of contract, it is difficult in such cases for either general or direct supervision of construction work to be provided by the manager of the project for the client or by any of the design team contributors. This responsibility is vested directly in the main contractor and subcontractors. Standard conditions of contract usually cast the architect or other manager in a passive role in connection with the construction work. The contract is directly between the client and the contractor and the rights and duties of both parties to the contract are specified. The architect or project manager monitors that the conditions of the contract are carried out. They often cannot intervene directly to ensure compliance with the contract but must follow the administrative procedures laid down, with final recourse to arbitration or law by either party to settle disputes.

If dissatisfied with the contractor's performance, and if satisfaction cannot be achieved by persuasion or mediation, the manager of the project for the client must recommend legal action to the client as a last resort. More purposeful management of the construction stage on behalf of the client will require forms of contract that differ significantly from those commonly used for competitive bidding. This will result in a consequent redistribution of risk and a realignment of responsibility for design and a subsequent redefinition of roles, as found, for example, in management contracting. Other initiatives have sought to tackle such problems in other ways, for example partnering which tries to change attitudes and aid collaboration, and design-and-build contracts which further emphasise the responsibility of the contractor and limit further the scope of the client's representative to manage the construction stage. Whichever approach is selected there remains the need to design an appropriate managing system which ensures that all the contributors respond to the needs of the client.

Pattern of managing system functions

The pattern of managing functions on a project will, therefore, be dependent upon the structure of the firms used in the project organisation and upon the client's organisation structure and the client's requirements regarding the approval powers it wishes to retain. The pattern will also depend to a large extent upon when the client introduces the project team into the process. However, the manager of the project on behalf of the client would normally undertake the activities of boundary control, monitoring and maintenance.

When a project organisation is designed, it is important that the people exercising the various managing activities are identified and their roles understood by all contributors. In this way the authority and responsibilities of the members of the contributing firms will be recognised. For example, it will be known whether the job quantity surveyor has full authority for quantity surveying matters or whether he or she is subject to direct oversight by a more senior member of the firm. This will depend upon the firm from which they come and their status within that firm.

The manager of the project team is usually involved in recommending courses of action to the client for approval. The manager's authority does not therefore generally derive from the power to approve the output of contributors but from the power of recommendation, which implies approval of the output, and hence the power to influence decisions made by the client. The manager's authority stems from access to the client and although this should not bar other contributors from the client if integration of the client is to take place, the latter can vest authority in the manager by considering recommendations only from this source and by requiring other contributors to route recommendations through that manager. Only in this way will the manager have the authority necessary to ensure that the other contributors perform adequately, and have the opportunity to exercise fully his or her integrating activities. Nevertheless, this situation will only be maintained for any length of time if the manager has the professional respect of the other contributors. The manager of the project may be under general supervision by another person higher in the hierarchy of the firm and this may affect the regard in which the manager is held by other contributors, at least initially. The manager's authority is likely to be enhanced if they are a partner or director of the firm.

It would be beneficial if the client were to state formally the authority of the manager of the project and of the other contributors. However, the informal authority of the manager, derived from the respect afforded by the client and other contributors, will be a potent factor and will be the instrument most likely to elicit the necessary level of performance from all contributors.

An example of how functions may be distributed is shown in Fig. 7.6. However, as has been stressed, there are many ways in which roles and functions may be distributed, and Fig. 7.6 shows in outline only one example.

The allocation of responsibility for the project among the contributors will depend upon the association of firms involved and will be the subject of negotiation between the client and the contributors. However, it is possible to put forward ideas as to how responsibility may be distributed. Conditions of contract for construction work will usually define responsibility for this aspect and the related responsibility of the other contributors in connection with this stage, but they do not have to follow the standard format and can be tailored to suit particular projects. Responsibility for design and associated work is the aspect for which responsibility can be more difficult to define.

If a project is managed and designed by a multidisciplinary practice, then responsibility will rest with that firm. Similarly, responsibilities when a conventional arrangement is used, with the architect as designer/manager, are generally understood. In this arrangement, if the consultants are directly appointed by the client, then they will be responsible to the client for their own work. The difficulties that arise in this respect result from the interrelationship of the contributions made and hence from the allocation of final responsibility for specific deficiencies.

If management of the project for the client is given to a firm separate from the firms making up the operating system, a comparable situation will arise if they are appointed directly by the client. Alternatively, if the project is managed by a firm which appoints the consultants directly, i.e. as 'subcontractors' to them, then the managing firm will take responsibility for their work and hence the total project. If a legal action is successfully brought against them by a client, they may have recourse against their 'subcontractors'. This argument can be extended to design-and-build contracts in which one firm will be responsible for the whole of the design and construction of a project.

Naturally, the greater the responsibility accepted by a firm the greater is the risk they are carrying. Firms, whether managing or design-and-build, are unlikely to accept higher responsibility unless they have direct control over the contributors through direct employment, or a facility to bring an action against a contributor

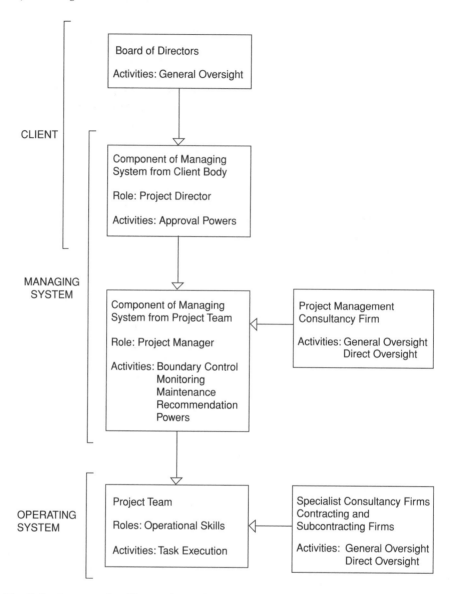

Fig. 7.6   An example of how roles and activities may be distributed.

if one is brought against them by the client. Importantly, their profit margin would have to be sufficiently high to allow for the level of risk being accepted.

A situation in which responsibility for the project rests with only one firm, or at least one firm for management, design and related aspects and one firm for construction, is likely to be attractive to clients. Informed clients are in a position to dictate the pattern that they want for their project although the reaction of the contributors is likely to be to wish to spread responsibility unless they are appropriately recompensed for risk. Ideally, responsibility should be matched by authority, but this is particularly difficult to achieve. The responsibility pattern adopted should reflect the project structure and the approval and recommendation pattern required by the client. It also should provide the client with legal protection that is sufficiently practical to be applied.

Project management activities

The abstract account of the functions of the managing system given previously needs interpreting into the activities that should be expected from project management. In undertaking this, concentration is placed upon project management carried out by the manager of the project team rather than by the client's components of the managing system. This is not to minimise the management of the project that takes place within the client's organisation. It is fundamental to the success of the project and provides the context which to a large extent determines how effective the project team can be.

The client should involve the project team at as early a stage as possible and, prior to involving them, should bring forward the initial need for the project in a co-ordinated and controlled way from within its organisation. If the project team is not involved in the early stages of the project, ideally, the client should have taken all useful advice from within its organisation, and should have identified all stakeholders and their interests, before bringing forward for the project team's advice the strategies that it believes will fulfil the objectives. The client should not have too rigid ideas at this stage as advice from the project team could lead to a better solution to the problem. It would be advantageous if the person who has co-ordinated the client's work and brought forward the strategies could form the client's component of the managing system when the project team is introduced.

The way in which the client carries out this process will be determined by the structure of its organisation but it is important that it should be made explicit to the person from the project team who forms their component of the managing system (referred to in what follows as the project manager). As mentioned previously, the person managing the project team for the client does not have to carry the title of project manager and it may be the architect or engineer who is responsible for managing the process; however, the term is used for convenience.

Consequently, the activities of project management that are discussed here are those that follow from this point and are carried out by the project manager. The detailed activities identified will not necessarily be performed on every project but are intended to illustrate the breadth of project management activities that may be required. Many of them will be repeated in finer degrees of detail at the different stages of the project as further information becomes available. They have not therefore been allocated to the major systems of the process, but require interpretation by the project manager in the light of their degree of applicability at the various stages. The order in which the activities are listed is not intended to indicate a sequence, as many of the functions are interdependent and will overlap in practice.

(1) *Establishment of the client's objectives and priorities*
If the project manager is involved in the conceptual stages of the project, this will be the starting point and it should encompass the client's and stakeholders' broader organisational objectives as well as the objectives for the envisaged project. This will enable the project manager to provide informed advice upon the alternative construction strategies available and so provide the basis for developing an appropriate brief for the project.

Even if the project manager is not involved until after the client has arrived at its own conclusions and has some firm ideas about the brief, the client's and stakeholders' broader objectives should be explored. This is to make sure that the client's ideas of what the project will provide are sound and are likely to satisfy those objectives.

(2) *Advise the client of the impact of environmental forces*
As part of the project manager's involvement at the conceptual stage, the project manager should alert the client to the effect on the project of any features of the project's environment of which the client may not be aware. Changes in regulation of the construction industry, and legal cases affecting contracts, are examples. A relatively recent but increasing environmental force is the effect of sustainability on the construction industry. Taking environmental (a confusing dual use of the word!) issues into consideration early (whether an environmental impact assessment is required or not) can be beneficial by producing a project which is defensible in the public domain; but a client may not be aware of this. The extent to which the client wishes to impose environmental requirements on the contractor could also be relevant.

(3) *Design of the project organisation structure*
The design of the organisation structure of the project should commence by anticipating the decision points in the process and defining the feedback loops and the relationships of the contributors to each other and to the decision points, as described in the last chapter.

(4) *Identification of the way in which the client is integrated into the project*
This will arise from the design of the organisation referred to in (3) above but merits a separate reference. It is important that the project manager persuades the client that the organisation has to be designed to mesh with the project team. Having achieved that, the project manager must ensure that the client responds to the need to integrate with the project team. This will take place at a formal level through meetings but the project manager should seek to ensure that discussions, decisions and the need for those decisions are passed through the appropriate channels of the client's organisation. This will require the project manager to adopt a position close to the client's organisation, and is important for the transmission of information to the client and for the project manager to sense and follow up any changes in the environmental context of the client's organisation that may affect the project.

(5) *Advice on the selection and appointment of the contributors to the project and the establishment of their terms of reference*

The client may require to be advised on these matters. It may even be the case in some circumstances that the client will leave this entirely up to the project manager. It is important that the contributors used on the project have the experience and capability for the particular project and that their approach is compatible with that of the project manager. The project manager is more likely to be able to make this judgement than the client unless the latter is very experienced in construction. The actual terms of appointment of the contributors will, of course, be a decision taken by the client but the project manager should be able to acquaint the client with the alternative approaches available and advise on that which is most suitable for the particular project.

This area also includes the terms of appointment of the project manager and the extent of the project manager's authority. The project manager can hardly advise on whom to appoint as project manager (!), but will have to negotiate the appointment with the client. Perhaps the most difficult decisions the client will have to make are whom to appoint to manage the project and, secondly, how to form an integration with the project manager from within his or her own organisation. The degree of authority delegated to the project manager will depend upon the extent to which the client wishes to retain power of approval, as discussed previously.

(6) *Translation of the client's objectives into a brief for the project team and its transmission*

This involves establishment of user needs, the budget, cost and investment plans and, where appropriate, disposal strategies and their correlation. The extent to which stakeholder interests are to be taken into account needs to be decided. It is at this stage that fundamental misinterpretations occur and opportunities for economies are overlooked which then become enshrined with the development of the brief. Construction of the brief requires considerable knowledge of the performance in use of buildings, from the points of view of value, maintenance and user activity, and the widest spectrum of advice should be obtained and the most appropriate techniques of evaluation of alternatives employed.

The project manager should make sure that the brief is clearly transmitted to other members of the project team and that it is understood by them. There is a perpetual danger that it will be misinterpreted and result in contributors pulling in different directions. This makes it particularly necessary that it is unequivocally drafted.

(7) *Preparation of the programme for the project*

Although this can be thought of as being part of the brief, it is identified separately since it is an area much neglected, particularly during the early phases of the project and the design stage. The programme should represent a realistic co-ordinated plan of the time needed for the project from the start until, and including, commissioning. As for any plan, it must be carefully monitored, controlled and adapted as necessary.

Although a number of useful techniques exist for programming (e.g. networking and bar charts), for building work none of the professions involved in the early stages of projects tends to specialise in this work and it is often left to the project manager to provide this operational skill. The situation tends to be different for civil engineering projects for which the engineer will often prepare the programme.

(8) *Activation of the framework of relationships established for the contributors*

Having established the relationships necessary for the project when designing the organisation and agreeing the terms of reference of each contributor, the project manager should ensure that the relationships and responsibilities are activated in the manner intended. The project manager therefore has to be close to the activities of the contributors to ensure that they are in fact performing the work allocated to them and are consulting and taking advice as intended. This activity also covers the work and contribution of the client.

(9) *Establishment of an appropriate information and communication structure*

Much of the information produced for projects is unco-ordinated and this can lead to inefficiency in information use and communication, and to misunderstanding. The project manager should lay down formal communication channels and determine the way in which information is to be presented. This is a particularly difficult problem but the Co-ordinated Project Information system will help and project management firms have devised and developed their own computer-aided data co-ordination systems. In the absence of such sophisticated systems, there is still the opportunity to require the project team to produce information in a compatible manner.

Communication channels present a less intractable problem and the project manager should design them to ensure that all relevant parties are kept up to date formally on events on the project. Of special importance in this respect is communication with the client. The responsibility for preparing co-ordinated reports for the client will rest with the project manager, who should agree the need and frequency of such reports with the client when designing the integrating mechanism between the client and the project team.

(10) *Convening and chairing meetings of appropriate contributors at all stages*

This represents the formal aspects of the integration of the contributors alongside which informal integration will be taking place. Such meetings will act as formal checks on the achievement of the brief in terms of design, cost and time. As such they must be action-minuted so that control can be exercised against the decisions of previous meetings. Naturally, minutes should be circulated to all contributors.

It is important that the project manager should be conscious of the need for meetings and adjust their frequency to suit the particular stage of development of the project.

(11) *Monitoring and controlling feasibility studies, design and production to ensure that the brief is being satisfied, including adherence to the budget, investment and programme plans*

This is the 'meat' of the project manager's work and is predominantly concerned with control. The project manager is responsible for taking appropriate action to ensure that the project proceeds to plan. The project manager should be prepared to advise the client if its requirements cannot be met or if an alternative strategy to that contained in the brief emerges as more appropriate to the client's needs.

The project manager will be activating the feedback loops built into the process and measuring progress against the project's objectives, monitoring the project's environment and responding as necessary. The project manager will not therefore be concerned only with the state of the project's development at the time a feedback sample is taken but must also be concerned with forecasting events in the future to anticipate potential problems and attempt to resolve them before they arrive.

Clients generally do not feel that they are well informed about their projects and the responsibility for this lies with the project manager. Thus, as well as involving the client closely in the project, as previously discussed, the project manager should keep the client formally up to date on forecasts of the project team's performance so that action can be taken by the client in advance of a forecast event that may affect its organisation, e.g. delay in completion of the project.

The individual tasks will vary from project to project, but the types of task over which the project manager will have to exercise control and co-ordination are broadly as follows:

- Land acquisition
- Applications for planning consent
- Outline design strategies
- Budget and investment strategies
- Advice on finance, taxation and grants
- Detailed design
- Design cost control
- Disposal strategies
- Proposals for contractual arrangements for construction
- Appointment of the contractor
- Appointment of subcontractors
- Construction
- Cost control during construction
- Disposal.

In integrating and controlling the contributors undertaking tasks such as those indicated above, the project manager should ensure that all appropriate contributors are involved in each task and that the outputs of the tasks are compatible with each other and with the project objectives in terms of design, quality, cost and time. In order to achieve this, the project manager needs to be assured that the contributors are maintaining an appropriate level of progress in carrying out the tasks and are employing suitable techniques. In addition, the project manager will need to be satisfied that the contributors are using an appropriate number of staff of the right calibre and experience. For this purpose appropriate relationships with the contributors will need to be developed and they will have to accept the project manager's authority to be satisfied on these issues. If the project manager is dissatisfied, authority will be required to ensure that the contributors respond to any reservation held in this respect. This represents a sensitive area and although project managers should be able to achieve their requirements by informal means, they may on occasion have to fall back on an authority given to them by the client.

(12) *Contribution to primary and key decisions and to making operational decisions*

The project manager will bring forward to the client alternative proposals upon which primary and key decisions, as discussed previously, will be based and will assist the client in coming to the decision that best satisfies the objectives. Alternatively, the process may have brought forward a single recommendation. The project manager will undertake the presentation to the client.

Within the process, operational decisions, as discussed previously, will have to be made as a result of the activities of the contributors and it should be the project manager's responsibility to make such decisions.

(13) *Recommendation and control of the implementation of a strategy for disposal or management of the completed project, including commissioning the building and advising on arrangements for running and maintaining it when completed*

This includes the sale or letting of the completed project where appropriate or the 'putting into use' of projects that will be occupied by the owner. In the latter case, maintenance manuals should be provided and 'taught' to the client's personnel.

In commissioning the project, the project manager should involve all the contributors appropriate to this activity so that they can explain to the client's personnel how the project and its services are intended to be used.

(14) *Evaluation of the outcome of the project against its objectives and against interim reports including advice on future strategies*

This represents the final feedback loop and should provide information on the performance of the project team and the client. The distillation of experience on the project should assist the project team and the client to improve their performance on other projects.

Jawahar-Nesan and Price (1997) have identified best practice for effectively performing the list of tasks they identify as those to be undertaken by the owner/ owner's representative (project manager). Their list of tasks overlaps substantially with those given above as do the more recent typical terms of engagement for project managers published by the Chartered

Fig. 7.7   Skills required by a project manager, as identified by CIC (2001).

Institute of Building (2002) as part of their *Code of Practice for Project Management for Construction and Development.*

Project management skills

In order to fulfil the roles and carry out the functions required by project management, a project manager needs a demanding range of skills. The skills needed are well expressed by the Construction Industry Council (CIC) in their publication *Construction Project Management Skills* (2001). The skills are grouped into five topic areas and classified as to whether a project manager should have competence in a skill or needs only to have an awareness of it. In addition to a project manager's skills, the publication incorporates the skills of design managers, construction managers and trade contractor managers. Figure 7.7 shows the skills that CIC believe are required for project managers. The skills identified are self-explanatory; the CIC publication gives detailed descriptions. Whilst the identification of the skills required is comprehensive and useful, classifying 'Organisation and People Skills' as requiring only awareness on the part of a project manager is surely an anomaly. Organisation and people skills are fundamental to project management, so require competence in them on the part of the project manager.

Some practical considerations

The Latham Report (Latham 1994) recognises that, whilst every project has to be managed, a separate or external project manager may not be required as many clients have their own in-house project management capability, which will probably be the most satisfactory arrangement. Alternatively it sees an external project manager being seconded into the client's organisation. But for the situation in which it is decided to retain an external project manager a series of recommendations are made. These centre on the need to:

- clearly define the appointment and duties of the project manager
- give the project manager the necessary authority
- ensure that fee levels for consultants are sufficient to provide the service required
- interlock the terms of engagement of all consultants.

These are important issues underpinning the effective exercise of the functions previously described. A further major practice matter which is rarely mentioned is the problem of continuity of the project manager. The Latham Report argues for a 'single person/firm', i.e. project manager, to pull the whole process together for the client but recognises that managers have been and still are often appointed for each of the stages of the project, e.g. design, construction. The discontinuity which this approach induces can have serious ramifications for the successful outcome of the project but, even when a project manager is appointed to cover the whole project, the disruption caused by changes in the project manager can have equally deleterious effects on the project outcome. Roberts (1989) believes that when a project has a succession of project managers the project is virtually always adversely affected. As changes of project managers cannot be eliminated, he identified a number of lessons for the incoming project manager:

- 'overlap at a milestone whilst the previous manager is still in post
- obtain a list of key contacts and rely less on the project files
- meet key people quickly to clarify key issues
- assume prior decisions were correct
- honour previous commitments
- monitor non-critical path activities, people will have been pursuing diligently those on the critical path but not the others
- if there is an overlap of project managers ensure the outgoing project manager deals with all the less significant administration before he goes.'

He concludes with:

'A final thought that I always keep in mind while in the midst of managing my projects is that I may not finish the project. I know the actions and inactions of my predecessors that caused me the most trouble. I document my files, fulfil promises, update the data base and take care of the non-productive work so that I can pass on projects that truly are clean and without problems.'

These concerns are reflected in Chapman's (1999) identification of the risk involved in changes in project personnel referred to earlier.

An associated problem, identified by Latham (1994) as a reason why there are mixed views about the benefit of project managers in the construction industry in the UK, is that project managers are often appointed after contracts 'run into difficulties or work has already begun and various procurement route options have thereby been blocked off'.

This chapter has intentionally focused on what the project management process should do, not on how project managers (or other titled persons) do it. A later chapter on leadership deals with much about how project managers do their jobs but there is a general level of understanding necessary prior to that, which is briefly discussed here.

The traditional statement of how managers go about their work, arising from the classical approach to management, is that they plan, organise, co-ordinate and control. The great challenge to this view came from Mintzberg (1973) when he asked but what do managers really do? Mintzberg considered that, at best, the classical view indicated some vague objectives managers have when they work. He set about destroying what he felt were four major myths about the manager's job, using the results of systematic research on how managers spend their time.

In terms of planning he found that managers are orientated to action and averse to reflective planning and that their activities are discontinuous with little time to settle on major issues. He found that a manager's work involves regular symbolic duties which cannot be delegated such as being present at ceremonies and negotiations. Also that managers strongly favour verbal communications and that formal information systems play little part in their thinking. He also challenged the view that management was quickly becoming a science and a profession, believing instead that managers rely on judgement and intuition. Instead of what he termed the folklore of management, he identified the basic roles of figurehead, leader, liaisor, environmental monitor, disseminator of information, spokesperson, entrepreneur, disturbance handler and resource allocator. Mintzberg believes that these roles are not easily separable, that they form an integrated whole in which no role can be removed from the framework and the job left intact. Whilst it is not the purpose of this book to develop these ideas in detail, for which the reader is referred to Mintzberg's work, this brief outline is important to provide the backcloth of what project managers (or others) actually do as they carry out project management activities.

The roles identified by Mintzberg and the previous formal statement of the activities of project management both subsume what is probably the most important role of the project manager, that of integrator. If the project manager were unable to fulfil such a role, it is extremely doubtful whether the functions suggested could be carried out satisfactorily.

The list of activities tends to stress the inanimate parts of projects yet one of the most important components is understanding the human aspects of projects, which can only be achieved by working through people. A vitally important part of the project manager's work is therefore concerned with listening and talking to the members of the project team. This will enable the project manager to anticipate problems that lie at the interface of the work of the contributors, and together with them come up with solutions.

A particularly difficult area in this respect is the project manager's relationship with the contractor and subcontractors when the project has been let by competition. The standard forms and conditions of contract often used in such situations tend, because of the formality and financial implications of the conditions, to inhibit informal relationships between the project manager and design team members, and the contractor and subcontractors. Nevertheless, because of the significance of the construction stage to project success, the project manager needs to be able to establish with the contractor the particular problems anticipated in the construction programme and the action necessary to overcome them. This refers in particular to the transmission of information from the designers to the contractors. It could be the case that, by using other methods of appointing the contractor and subcontractors that allow them to be involved in the design stage, these types of problem will, to a large extent, be overcome.

An essential support to project management activities is the ability of the project manager to understand other people, to identify what makes them tick and hence to be able to motivate them to perform to the limit of their capabilities. The ability to do this arises from the personal characteristics of the project manager.

Finally, and failing all else, the project manager will need to 'arbitrate' in the case of formalised disputes on the project, whether within the design team, with the construction team or between them, in order to safeguard the client's position. The project manager's knowledge that at some stage such a position may need to be taken will make it more difficult to adopt the integrating role, which is of a more conciliatory nature, but the resolution of such conflict is what is expected of project managers.

Design of organisation structures

The ability of a managing system to operate effectively depends upon an appropriately structured operating system and complimentary managing system. The model has identified, in systems terms, the elements of importance in structuring organisations and has attempted to relate them, in abstract terms, to the construction process. It does not, therefore, present a rigid proposition for the organisation structure of the construction process, but proposes an approach that responds to the specific demands of individual projects. A role of the managing system is to design the organisation through which it will work in seeking to achieve the client's objectives. The manager must, therefore, be provided with the authority to design the operating and managing systems, and to make them function. Such authority will stem from the client, who must decide the pattern of authority established for the project.

Against the background of the model, and accepting that the client's objectives will have been spelled out, an approach to designing an organisation structure for a building project could be:

(a) The manager convinces the client of the need to design an organisation structure for both the construction process and for the client's own organisation's relationship with it, and makes the client aware of the demands that will be placed upon the client's organisation.
(b) The project organisation structure is designed at the very beginning of the process.
(c) Primary decision points are identified.
(d) Within each system created by the primary decision points, the key decision points are anticipated as far as possible.

(e) Within each main subsystem created by the key decision points, operational decision points are identified.
(f) Feedback loops are established within the structure produced by (c), (d), and (e).
(g) The task subsystems within the organisational subsystems are identified, together with the skills required to undertake them.
(h) The manner by which the skills are to be provided is established, e.g. separate specialist firms, design-and-build, etc., and allocated to the operating system.
(i) Methods of achieving the required level of integration needed because of the differentiation generated by the system are established, including methods of integrating the client.
(j) The pattern of managing system activities is identified as a product of the structure of the operating system and the manner by which skills are provided.
(k) The managing system functions of approval, recommendation, boundary control, monitoring, maintenance and general and direct oversight are allocated, which should include the authority and responsibility pattern of the managing system, client and contributors to the operating system.
(l) Methods of monitoring, mitigating and harnessing the effects of environmental forces are identified.

The extent to which this approach can be achieved in practice will depend upon the relative certainty of the climate in which the project is being undertaken. With a high level of uncertainty, it may be that the organisation can be designed only a stage at a time, but by going through the process suggested, areas of uncertainty will be recognised. For projects with more stable environments, the organisation structure will be easier to lay down and should present a sound basis for close control of the project.

The essence of the approach is that one starts with a statement of what is to be accomplished through identification of the decision points in the process. Bennett (1991) believes that experienced practitioners can predict key decision points for project types with which they are familiar and at the very least can predict the work required to reach the next key decision point for less familiar projects as there are distinct families of projects with similar sequences of decision points. Identification of decision points is then followed by the design of an operating system required to undertake the tasks prior to each decision, and only then is the managing system designed to suit what is to be achieved. Thus, organisation structure design *follows* the process flow rather than the process having to fit into a predetermined operating and managing system.

Chapter 8
Authority, Power and Politics

Introduction

Authority has been referred to on a number of occasions and in practically every case reference has been to formal authority as the result of a structural position within a project organisation. This simple approach to authority was necessary in order not to deflect attention from the main point being made at that time to which authority was then not central. But now is the time to address the concept of authority in organisations in much more depth and more particularly its more complex companion power and its bedfellow organisational politics. All are strong forces in organisations, be they corporate or project organisations. No matter how well the organisation structure of a project is designed to suit its task and environment, the project will be unsuccessful if the authority pattern is inappropriate and if the powerful players and their political agenda are set against the project. It is worth repeating here that the significance of power in project teams is the opportunity of the person in a position of power to influence the objectives of the project and hence the satisfaction of the client with the project's outcome. The relationship between authority and power needs to be understood but first a development of the concept of authority used so far is necessary.

Authority

Authority is intrinsic in achieving objectives through organisations. Authority which is vested in a person through the position held in the organisation, and hence the person's right to make decisions upon which others are required to act, is the most common understanding of authority in organisations. Such authority is therefore seen as essential in order to get things done. This simple view of authority describes *formal* or *legal* authority and rests on three basic assumptions (Cleland & King 1972):

- 'the organisation chart is a realistic descriptive model of an organisation.
- legal (or line) authority is delegated down through the "chain of command".
  Therefore, if one has legal authority, one can demand the obedience of others.

- given sufficient authority, an individual can accomplish organisational objectives regardless of the complexity of the forces that are involved (Ries 1964).'

But the concept of formal authority is insufficient in today's complex organisations and informal patterns overlay the formal structure. But before moving on to examine such phenomena, the origins of formal authority are worth exploring.

The original source of authority in society can be traced back to private property rights which were held by the Crown and the Church. They owned the land from which food was produced and hence had the power to enforce their authority. This was the model from which authority in industrial processes arose during the industrial revolution. Independently of the rise of this aspect of traditional management 'theory', Max Weber, the influential German sociologist, produced his typology, (summarised by Scott (1992)) which identifies three types of authority (Weber 1968 trans.):

Traditional authority – resting on an established belief in the sanctity of immemorial traditions and the legitimacy of those exercising authority under them.
Rational–legal authority – resting on a belief in the "legality" of patterns of normative rules and the right of those elevated to authority under such rules to issue commands.
Charismatic authority – resting on devotion to the specific and exceptional sanctity, heroism or exemplary character of an individual person, and of the normative patterns or order revealed or ordained by him or her.'

Traditional authority has its roots of feudalism; rational–legal authority underpins Weber's notion of bureaucracy. He believed the charismatic forms arose in periods of instability and crisis when people turned to people they believed (rightly or wrongly) could resolve crises. Thus to a large extent a formal view of authority was enshrined within early management thinking.

The work of Barnard (1938) took the understanding of authority significantly forward. His view that goals are imposed from above whilst their achievement depends on willing co-operation from those lower down the organisation, led to his view of authority in which he states that it is a 'fiction that authority comes from above' and that 'the decision as to whether an order has authority or not lies with the person to whom it is addressed and does not reside in persons of authority or those who issue these orders'. These ideas gave root to the concept of *informal* authority which is now explicitly recognised as transcending formal authority and is subsumed within the concept of power in organisations which will be discussed later.

These ideas were developed by Dornbusch and Scott (1975) who identified three types of authority: endorsed, authorised and collegial. (They actually referred to these as endorsed power, authorised power and collegial power but the use of power in this context is confusing as explained later when the differences between authority and power are explored.) Originating from Barnard's work they identify authority by endorsement as that arising from a subordinate group acting as a coalition which limit and regulate the exercise of authority over them by a superior. They then go on to recognise that an important characteristic of formal organisations is that each superior will (in the large majority of cases) also have a superior in the hierarchy. Subordinates will have the opportunity to appeal to their superior's superior with the hope that their immediate superior's authority will be curbed. They term the authority of the superior's superior as authority by authorisation. A further source of enforcement of authority norms is seen to be the colleagues or equals of the superior in question. They believe that this source may be expected to be of particular importance in professional organisations and is referred to as authority by collegiate.

Formal authority has evolved, but Leavitt (2005) believes it 'has been softened, even cloaked in cordiality. But it is still there, and it's still the central reality of organisational life'. Whilst formal (or legal) authority is modified by power structures much of the management literature tends to overstate the absolute influence of informal aspects. In business, people are always likely to respond to a command from their superior. However they may not carry out the order effectively or may seek to dodge it. This latter possibility is high when a subordinate's knowledge of his or her task is greater than that of the superior. Pfeffer (1992), in an account which demonstrates the strength of formal authority, states that 'obedience to authority is conditioned early in life and offers, under most circumstances, many advantages to both society and the individual'.

Whilst legal authority is given by the ability to impose sanctions, the greatest of which in a business sense is the withholding of a salary rise or ultimately dismissal, there are many constraints on legal authority. Formal limitations are contained in laws, contracts, etc., and informal limitations are provided by morality and the capacity of the person so ordered to carry out the task. In fact absolute legal authority does not exist as it depends upon the sanctions being sufficiently high to make the person so commanded obey the order. In a business sense an employee may accept the sanction of dismissal rather than obey the order. In a more extreme example, authority imposed by a state over its people may not be accepted and so revolutions begin and the ultimate sanction of death is not considered great enough to command obedience.

Power is the influence individuals have over the people with whom they interact to the extent that they do as they require. It operates with subordinates, superiors, peers and friends and has to be earned. If misused it can be counterproductive.

Cleland and King (1972) identify the following talents as contributing to what they term to an individual's informal authority but which is now usually referred to as power:

- 'superior knowledge
- an ability to persuade people to his way of thinking
- a suitable personality and the ability to establish rapport with others
- a favorable reputation with peers and associates
- a record of accomplishments which lends credence to his or her experience and reputation

- an ability to build confidence in peers and associates
- patience to listen to the problems of subordinates and peers and a willingness to help out when asked or when the need to help is sensed
- an ability to resolve conflict between peers, subordinates, and associates.'

The strength of a person in a position of high formal authority who also has these characteristics is not difficult to imagine, nor are the connections between effective leadership, formal authority and these characteristics.

Power

It is important to distinguish between authority and power in organisations. Power is a much broader concept than authority (Weihrich & Koontz 1993) and many pages have been devoted to its definition. Many stem from Weber's (1947 trans.) definition that 'power is the ability of a person to carry out his own will despite resistance'. Hence 'the ability of individuals or groups to induce or influence the beliefs or actions of other persons or groups' (Weihrich & Koontz 1993), 'the capacity of individual actors to exert their will' (Finkelstein 1992) and the simple definition 'the ability to get things done' (Kanter 1983) used by Lovell (1993) which he believes has a significant appeal in relation to the role of project managers.

Scott (1992) relies on the approach of Emerson (1962):

'It would appear that the power to control or influence the other resides in control over the things he values, which may range all the way from oil resources to ego-support, depending upon the relation in question. In short, power resides implicitly in the other's dependence.'

He believes that Emerson's approach means that power is not to be viewed as a characteristic of an individual but as the property of a social relation. The social relation is defined by dependency. When one person is dependent on another person, that person gains power over them (Robbins 2005). This means that in defining the power of an individual it is necessary to specify over whom he or she has power. That is, an individual cannot have power generally. Nevertheless, the possession of power is generalised by Pfeffer (1992) in defining power as 'the potential ability to influence behaviour, to change the course of events, to overcome resistance, and to get people to do things that they would not otherwise do'.

Pfeffer defines organisational politics as the exercise or use of power. That is, politics and influence are the processes, actions and behaviours through which this potential power is utilised and realised. Scott comes to a not dissimilar conclusion in combining the definition of power and its use as:

'We will define interpersonal power as the potential for influence that is based on one person's ability and willingness to sanction another person by manipulating rewards and punishments important to the other person.'

Whether such a narrow definition of power is relevant in a construction project management context is open to question and is discussed later.

Relationship between authority and power

Most dictionary definitions of authority contain a reference to power. For example:

'Legal power or right: power derived from office or character or prestige.' (*Chambers Twentieth Century Dictionary*)

We know that formal authority stems from a structural position in an organisation and can be enhanced by the characteristics of the person in that position. The person in that position will have power over his or her subordinates which can be conceived as legitimate authority or legitimate power. In this conception the terms authority and power can be seen to be synonymous.

The situation in which power gains its wider meaning is when it is used outside the legitimate (or formal) authority structure such that the person exercising power imposes his or her will on others in the organisation. Such power could be described as illegitimate power as it is used outside the formal organisation structure. For this reason there exists a general feeling of disquiet about its presence in business organisations (it appears much more acceptable in overt political systems.) This leads Pfeffer (1992) to believe that we are ambivalent about power and to quote Kanter (1983) who believes 'Power is America's last dirty word. It is easier to talk about money – and much easier to talk about sex – than it is to talk about power'.

However, in its illegitimate form power can be a 'positive' or 'negative' force. In its positive form those with power will be using it to further objectives which are directly compatible with the organisation's official objectives. In its negative form power will be used to achieve objectives which do not subscribe to organisational objectives. Even in what may appear to be its positive form it can have negative aspects as control of such forces is by definition difficult as they are not explicit and whilst on the face of it may appear positive they may in fact be exerting a negative effect.

Positive power has been described by French and Bell (1990) as 'a balanced pursuit of self-interest and interest in the welfare of others; viewing situations in win–win (non zero-sum) terms as much as possible; engaging in open problem solving and then moving to action and influencing'. They see negative power, on the other hand, as 'an extreme pursuit of self-interest; a tendency to view most situations in win–lose (zero-sum) terms; and predominant use of tactics such as secrecy, surprise, holding hidden agendas, withholding information or deceiving'.

Scott (1992) points out that authority structures are more stable and effective control systems than power structures. The authority structure not only allows a greater measure of control by the person in authority but equally, if not more so, the authority structure also regulates and defines that power. It allows those under authority to act as a coalition in relation to the person in authority to define the limits within which that power can be exercised.

The sources of power

There have been many perspectives on the sources of power in organisations, some of which are complementary, some of which overlap. Together they help to explain the nature of power in organisations.

Emerson (1962) sees power as relational, situational and potentially reciprocal. That is it occurs between people in specific circumstances and may work in both directions. The basis of his formulation is that the power of superiors is based on the sanctions they hold over others and their willingness to use them but that a reward or a penalty is determined by the goals or value of the subordinate in the relationship. In this approach power can have many foundations but writers often refer to power being based on the resources which can be employed in the attainment of desired goals. In this context resources are defined extremely broadly, for example money, skills, knowledge, strength, sex appeal. What types of resource will function as sanctions will vary from one individual to another and from situation to situation (Scott 1992).

Emerson considers that power relations can be reciprocal in that an individual may have power over another in one matter but power may work in the opposite direction between the same people in another matter. Such power relations are likely to be significant in highly interdependent areas of business such as those found in connection with construction projects.

Within this context there is a reasonably consistent view of the sources of power (Weihrich & Koontz 1993; Newcombe 1994) in the categories of reward, coercive, expert and referent power. Reward power refers to the power to offer enhancement (in many forms), for example, pay, position, conditions of service. Coercive power is closely related to reward power and is the power to punish by way of deprivation of benefits, for example pay or status. Whilst these categories are frequently seen to be legitimate and organisational and equivalent to formal authority, they do not have to be so. Many indirect opportunities to reward or punish exist outside the formal authority structure and can be seen as the exercise of illegitimate power.

Expert power arises from skill, knowledge and, increasingly, information. Finkelstein (1992) believes that the ability of top managers to deal with environmental contingencies and contribute to organisational success is an important source of power. Managers with relevant experience may have significant influence on a particular strategic choice and are often sought out for their advice. However, power tends to accrue when a manager's expertise is in an area critical to an organisation.

Referent power is the influence that people exercise because people believe in them and their ideas. Finkelstein refers to this as prestige power. He believes that managerial prestige promotes power by facilitating the absorption of uncertainty from the institutional environment both informationally and symbolically. He also believes that prestige also provides power through suggesting that a manager has gilt-edged qualifications and powerful friends.

Whilst these sources of power are accepted, there are attempts to subdivide and extend them. For example, Kakabadse *et al.* (2004) use reward power (which includes coercive power), role power (which is formal authority), personal power (referent power), knowledge power, network power and information power (which can all be seen to be part of expert power) and corporate memory power (which may also be seen as expert power).

These sources of power are seen as personal and although they can operate within the formal authority structure, and will have a great effect as confirming its legitimacy, they possess the ability to operate freely outside the legal authority structure and be potent sources of power which can undermine an organisation's objectives if used illegitimately. Scott's definition of power can now be seen to be narrowly drawn and focused on power within the legitimate authority structure of the organisation.

Personality or charisma are often referred to as a source of power (Robbins 2005) but are better perceived as a reinforcing agent to the sources described. The strength gained through expert or referent power will be increased by an appropriate personality.

Finkelstein's excellent paper also deals with structural power which is conceived as the equivalent of the formal authority structure. He also identifies ownership power which he defines as 'power accruing to managers in their capacity as agents acting on behalf of shareholders'. He believes that the strength of a manager's position in the agent–principal relationship determines ownership power. The strength of managers' ownership power depends on their ownership position as well as on their links to the founder of a firm so that managers who are also shareholders are more powerful than those who are not.

Power and interdependency

Pfeffer (1992) recognises the link between power and interdependency in organisations. He believes that power is used to different degrees depending on the level of interdependency. With little or no interdependency there is no need for power as there is no situation in which dependency occurs. Similarly he believes that when interdependency is high the motivation to work together is also high and that if this incentive is ignored the organisation is likely to fail. It is in conditions of moderate interdependence that Pfeffer believes that power is more frequently used. However, this view appears to be an oversimplification as it reflects a legal or formal use of power (authority). It is probably the case that generally low interdependency sees little use of power but even in such situations there may be people working on other agendas in which they use illegitimate power with a view to a pay-off in other circumstances in future. But this would be much more likely to occur in situations of high interdependency where the opportunity for the use of power is likely to be greater. This situation is one which could be frequently found on complex construction projects.

Interdependency means that dependency of one person or unit on others exists. It is this dependency which is the basis of power. If someone possesses something that another requires which is not available elsewhere, so that they alone control it, they gain power over the other (Robbins 2005). This is particularly so in the case of knowledge and information. The ability to diagnose the power structure in organisations is vital to achieving objectives (either legitimate or illegitimate) and, to do this, patterns of interdependency have to be understood. In project management terms this applies within both project teams and client bodies.

As stated by Liu *et al.* (2003), citing Walker and Newcombe (2000):

'A consequence of specialization in the construction industry is the dependencies of the activities and the necessary integration to obtain the desired output. Such integration leads to power structuring directly due to the dependencies. However, in a process like the realization of construction projects, such structuring is dynamic and so the shifting multi-goal coalition which results reflects the changing power structure of the main actors.'

The complexity of dependencies in the realisation of construction projects is such that if is often difficult to determine who has the greatest dependency at any given time: the project team members depending on the project manager, the project manager depending on the project team members, or the project team members among themselves.

Politics in organisations

Organisational politics is the exercise or use of power (Pfeffer 1992). Politics has to do with power, not structure (Mintzberg 1989). Whilst Mintzberg agrees that political activity is to be found in every organisation he believes that politics act to the detriment of the effective functioning of organisations by 'disordering and disintegrating what currently exists'. He unequivocally states that 'I am no fan of politics in organisations' but he recognises that no account of the forces at play in organisations can be complete without a consideration of politics.

Mintzberg uses an analogy of politics as organisational illness which needs to be understood. He sees it working both for and against the system. Politics can undermine healthy processes but can also strengthen a system by acting as a symptom of a more serious disease, enabling early action by provoking the system's adaptive mechanisms. Mintzberg identifies thirteen political games which by his definition involve illegitimate use of power but many of which use legitimate authority as a part of the play. Some co-exist with strong legitimate authority and could not exist without it, others, usually highly divisive games, are antagonistic to legitimate authority and further games arise when legitimate authority is weak and those games then substitute for it.

Two of the games will be more easily recognised by construction project management people:

'Expertise game: non-sanctioned use of expertise to build power base, either by flaunting it or by feigning it; true experts play by exploiting technical skills and knowledge, emphasising the uniqueness, criticality, and irreplaceability of the expertise, also by seeking to keep skills from being programmed, by keeping knowledge to selves; non-experts play by attempting to have their work viewed as expert, ideally to have it declared professional so they alone can control it.

Rival camps game: played to defeat a rival; typically occurs when alliance or empire-building games result in two major power blocs, giving rise to two-person, zero-sum game in place of n-person game; can be most divisive game of all; conflict can be between units (e.g., between marketing and production in manufacturing firm), between rival personalities, or between two competing missions (as in prisons split between custody and rehabilitation orientations).'

Other games are also relevant, particularly in terms of understanding politics in client organisations and the firms from which project team members come.

As referred to in Chapter 3, Mintzberg has approached the study of organisations on the basis of configurations and identified five basic types of organisations. One type is the political organisation which he describes as:

'What characterises the organisation dominated by politics is a lack of any of the forms of order found in conventional organisations. In other words, the organisation is best described in terms of power, not structure, and that power is exercised in ways not legitimate in conventional organisations. Thus, there is no preferred method of co-ordination, no single dominant part of the organisation, no clear type of decentralisation. Everything depends on the fluidity of informal power, marshalled to win individual issues.'

Of which he identifies four forms of political organisation:

- 'Confrontation: characterised by conflict that is intense, confined, and brief (unstable)
- Shaky alliance: characterised by conflict that is moderate, confined, and possibly enduring (relatively stable)
- Politicised organisation: characterised by conflict that is moderate, pervasive, and possibly enduring (relatively stable, so long as it is sustained by privileged position)
- Complete political arena: characterised by conflict that is intense, pervasive, and brief (unstable).'

The 'complete political arena' appears to be an extreme type of organisation probably rarely found in its absolute form whilst the other three are probably less uncommon but still rare.

The shaky alliance form is the one which is most easily recognised in construction project management terms. It exists when 'two or more major systems of influence or centres of power must co-exist in roughly equal balance'. Mintzberg uses the symphony orchestra to illustrate this form but those in construction will find it familiar, e.g. architect/engineer, designers/constructors.

The political organisation is not the only form subject to political influences which is why it was earlier described as extreme and rarely encountered in its ultimate state. As Mintzberg recognises, politics exist in his other organisation classifications to varying degrees but less extensively than is needed to classify them as political organisations.

Two of Mintzberg's organisation categories – professional and innovative – are particularly relevant to the management of construction projects. The professional organisation relies on the standardisation of skills, which is achieved primarily through formal training. It hires professionals for the operating core, then gives them considerable control over their own work. The innovative organisation relies on 'adhocracy'. Mintzberg believes that 'sophisticated innovation requires a very different configuration, one that is able to fuse experts drawn from different disciplines into smoothly functioning ad hoc project teams'. The uniqueness of construction project management organisations is that they often combine Mintzberg's professional and innovative configurations.

What is significant is that Mintzberg believes that there is considerable room for political games in each of these configurations as both have relatively weak systems of authority, though strong systems of expertise. When combined as in construction project management their ability to convert to political organisations must be even greater.

One can do no better than quote Mintzberg:

'The professional configuration may have a relatively stable operating core, where activities are highly standardised, but its administrative structure, where all kinds of professionals and managers interact to make choices, is hardly stable and, in fact, very supportive of power games. The innovative configuration is far less stable, generally having a highly fluid structure throughout that literally promotes games.'

Whilst taking a firmly negative view of politics in organisations Mintzberg does recognise that politics can serve a functional role:

- 'As a system of influence it can ensure that the strongest members of an organisation are brought into positions of leadership
- It can ensure that all sides of an issue are fully debated
- It can stimulate change that is blocked by legitimate systems of influence
- It can ease the path for the execution of decisions.'

Whilst reluctant to endorse any political organisation as functional he does acknowledge that a shaky alliance that reflects natural balanced and irreconcilable forces in the organisation could be functional. He uses the example of the differences between research and manufacturing people in a firm which needs the two in balance and considers the alliance functional in such a situation. This is a situation which is directly analogous to design and construction. He goes on to say 'This is because the organisation could not function if it did not accommodate each of these forces. It has no choice but to take the form of a shaky alliance. Some conflict is the inevitable consequence of getting its work done'.

Whilst Mintzberg's analysis and illustrations remain amongst the most cogent, there are many other accounts of politics in organisations that are valuable, but which mostly cover the same ground even though it has been held that 'writer, researchers and managers are not able to agree on the true meaning of the term politics' (Kakabadse *et al.* 2004). They describe the situation within organisations as a competition 'for resources, for attention, for influence; there are differences of opinion concerning priorities and objectives; clashes of values and beliefs occur frequently. All these factors lead to the formation of pressure groups, vested interests, cabals, personal rivalries, personality clashes, hidden deals and bonds of alliance'. Within this milieu, politics is a process to influence other individuals or groups toward one's own views. Although not manifesting themselves in the same way as within non-project organisations, these features do have an effect on construction project organisations.

Other views on politics stem from the strategic contingency theorists who argue that power is used to shape the structure of the organisation which reinforces the potential for future political struggle. These views (Hickson *et al.* 1971; Pfeffer 1981) seem less relevant to temporary management organisations than to permanent organisations where long-term positions are likely to pay political dividends.

Power and leadership

The concepts of power and leadership are, of course, closely related. The later chapter on leadership examines the characteristics needed in leaders and the circumstances in which different characteristics and leadership styles are appropriate. Leadership is directly associated with both formal authority and the legitimate use of power. The link between leadership and power is therefore that the legitimate power of the leader is the potential for influence held by the leader as a function of his or her formal authority and personal characteristics and talents.

This raises the issue of the relationship of illegitimate power which has been a principal focus of this chapter. If strong illegitimate power is being used then it is possible to conceive the idea of an illegitimate leader, that is a leader operating outside the formal authority structure. Whilst there may be some good served by such a situation in terms of compensating for weak or misdirected legitimate leadership, the potential for serious dysfunctional conflict is extremely high. Therefore leadership patterns and performance in organisations should always be considered within the context of the power structure both legitimate and illegitimate.

Empowerment

In dealing with power comprehensively it is necessary to touch on empowerment. At first glance it appears to be an elaboration of delegation but close examination shows that it is far more than that. Stacey *et al.* (2000) appear to see empowerment as integral to complexity theory by stating that self-organisation is another term for empowerment, but then cloud the issue by stating that empowerment is 'really a more idealistic word for delegation'. The literature seems reluctant to provide a clear definition but Neilsen's (1986) is useful:

> 'Empowerment is giving subordinates the resources, both psychological and technical, to discover the varieties of power they themselves have and/or accumulate, and therefore which they can use on another's behalf.'

It can be seen that empowerment revolves around the need to provide psychological support and stems from an understanding of powerlessness. According to Kanter (1977):

> 'People held accountable for the results produced by others, whose formal role gives them the right to command but who lack informal political influence, access to resources, outside status, sponsorship, or mobility prospects, are rendered powerless in the organisation ...'.

People who feel powerless do not respond well in the organisational setting, do not provide the level of effectiveness required and often have a negative effect on the achievement of organisational goals. To overcome the negative effects of powerlessness a positive policy of empowerment is necessary. Burke (1986) provides five approaches which allow managers and leaders (between which he draws a distinction) to empower their followers and subordinates as shown in Table 8.1.

Conger and Kanungo (1988) identify leadership practices which are empowering and which are consistent with those of Kanter. They include:

Table 8.1 Differences in the empowering process as a function of role: leaders compared with managers (Burke 1986).

Empowering process	Leaders	Managers
Providing direction for followers/subordinates	Via ideals, vision, a higher purpose, superordinate goals	Via involvement of subordinates in determining paths towards goal accomplishment
Stimulating followers/subordinates	With ideas	With action; things to accomplish
Rewarding followers/subordinates	Informal; personal recognition	Formal; incentive systems
Developing followers/subordinates	By inspiring them to do more than they thought they could do	By involving them in important decision-making activities and providing feedback for potential learning
Appealing to followers'/subordinates' needs	Appeal to needs of followership and dependency	Appeal to needs for autonomy and independence

'the expression of confidence in subordinates including giving positive emotional support during experiences associated with stress and anxiety, the fostering of opportunities for team members to participate in decision making, the provision of autonomy free from bureaucratic constraint, the observation of others' effectiveness, i.e. providing models of success with which people identify, the setting of inspirational and/or meaningful goals, and, above all, the establishment of a trusting and cooperative culture.'

The benefits of empowerment are claimed to be that it motivates people to face greater challenges that they would if they felt powerless. People are likely to accept higher performance goals and hence leaders are able to put such challenges before subordinates with a reasonable expectation that they will respond. People will be motivated to persist in the face of difficulties due to their increased level of confidence in their own abilities to influence events. In a study of 111 work teams, Kirkman and Rosen (1999) found that, 'More empowered work teams were . . . more productive and proactive than less empowered teams and had higher levels of customer service, job satisfaction and organisational and team commitment'. Nevertheless, Conger and Kanungo believe that empowerment could have negative effects. They believe that empowerment might lead to overconfidence and hence misjudgements on the part of subordinates. Because of false confidence in positive outcomes, organisations may persist in efforts which are inappropriate to their aims.

Liu *et al.* (2003) identify the situation in which some leaders may feel threatened by empowered subordinates because there is some loss of control, and the corollary of some subordinates not wanting to be empowered as they are non-assertive and cannot accept responsibility. The question of responsibility does not seem to be clearly addressed in the literature on empowerment. A key issue would appear to be to ensure that those who are empowered also accept responsibility for their actions otherwise the false confidence referred to by Conger and Kanungo could lead organisations into unwise commitments.

Power in project management

Little has been written about power and political structures, either legitimate or illegitimate, in the project management process yet their use and their potential for influencing the outcome of projects is well recognised by those involved in project management. In his paper on sociological paradigms applied to client briefing, Green (1994) uses a political metaphor to explain the nature of multi-faceted clients. Newcombe (1994) applies a power paradigm to the procurement of construction work contrasting the nature of power in traditional and construction management systems of procurement, Poirot (1991) examined a matrix system and found it to be a power-sharing, power-balancing organisation and Bresnen's (1991) paper purported to show how goal and power differentials can affect project outcomes but, whilst describing how complex construction project organisations are, did not explicitly analyse the power structure.

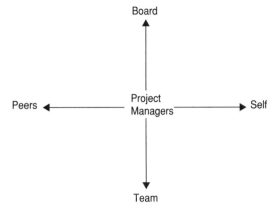

Fig. 8.1 Power and the project manager.
(Reprinted from *International Journal of Project Management,* **11**, Lovell, R.J. 'Power and the project manager', Fig. 1, p. 74, Copyright 1993, with kind permission from Elsevier Science Ltd, The Boulevard, Langford Lane, Kidlington OX5 1GB, UK.)

A comprehensive view of power in projects is given by Lovell (1993) who provides a model of how power affects project managers as shown in Fig 8.1. Lovell is discussing project managers generally and not specifically project managers in construction, hence his use of the term 'Board' in his model. This highlights the distinctive nature of construction project management relative to the more generic concept. The major feature in construction project management, which relates to power structures, is the degree to which the project is handled in-house by the client organisation.

At one end of the spectrum the project may be totally in-house including all design and construction capabilities, for example for a large developer or a petrochemical company. In such circumstances the potential for the illegitimate use of power would be high as political structures are capable of being established through the long-term relationships which will be present. At the other extreme is the small company which builds very infrequently and hence employs external consultants and perhaps a project manager. In such situations both legitimate and illegitimate power will be of a different kind and scale. Between these two extremes are a whole range of different configurations of client, consultants and contractor. Each project organisation structure will generate different legitimate and illegitimate power structures and the ideas relating to power discussed previously will need to be applied to analyse the appropriateness and effectiveness of the use of power in each case.

Formal authority in projects

The extent to which formal authority is explicitly stated in project structure varies considerably both between projects and within projects. The ultimate in explicitness is contained in the contract for construction work which lays down the rights and obligations of all parties. However, in the same project it is possible to have the project manager appointed by a brief letter with no explicit statement of either his or her authority or responsibility. Similarly the authority of the client over the project team is invariably implicit. On the other hand the project manager and the consultants may be appointed using formal agreements which are interrelated to show the authority and responsibility of all the contributors.

The formal authority structure within the client organisation itself is of importance to the project team as it will have a significant effect on the establishment of the brief for the project and will determine in the first instance who the project team should be listening to.

The ideas of endorsed, authorised and collegial authority referred to earlier are a useful way of thinking about formal authority in projects. The project team can be seen as a coalition which limits the exercise of formal authority over it by the project manager reflecting endorsed authority. This is due to their collective technical and professional knowledge which the project manager will have to take into account before issuing an instruction with which they may collectively disagree. Authority by authorisation can be seen as the project team's ability to appeal over the head of the project manager to the client. Although unlikely to be used, the fact that the opportunity exists, particularly for members of the team with strong illegitimate power, can limit the project manager's absolute authority. Collegial authority is that which operates between equal members of the project team. It is a powerful force in determining which members, and hence specialist contributions, are likely to hold sway when differences in opinion occur.

An issue on construction projects which differs from other organisations is the availability of sanctions. If a project is wholly in-house then the sanctions available to enforce formal authority will be similar to those in other organisations. However, when elements of the project team are external to the client organisation the available sanctions are different in nature, or at least the likelihood of them being used is remote. It would be a dramatic and potentially disastrous course of action to fire any of the consultancy firms. Whilst this has occurred on some projects it would be an absolute last resort. Intermediate sanctions are not readily available other than a threat to withhold future commissions.

Power in projects

Project teams are coalitions of highly skilled professionals in which opinions held about the project manager and the other professionals in the team have a profound effect on the power vested in each person in the team. This effect is influenced strongly by sentience, often resulting in the degree of formal authority being substantially weakened. Naturally these forces are particularly significant in terms of project success if they undermine the authority of the project manager. To stand the greatest chance of success the project manager needs strong formal authority, reinforced by the power to influence events.

The influence of the client on the formal authority of the project manager and the power of team members must not be overlooked. Although the client is the highest level of formal authority, any opinion which the client makes explicit about any member of the project team, particularly the project manager, will have a profound effect on that individual's influence within the project team.

The meaning of power used here is its use outside the legitimate or formal authority structure of the project and, as described earlier, arises from the dependencies between project team members, including the project manager. Also as previously stated, its potential for use will be much greater on wholly in-house projects and may well operate in such situations in ways which are comparable with the way in which it operates in organisations generally, as described in the mainstream management literature on power. It is in circumstances where projects are not carried out wholly in-house where special features of power arise specifically related to the organisation structure of construction projects.

The exercise of power in client organisations can have major effects on a project's objectives and how they are defined. Power struggles within client organisations can distort objectives leading to the construction of a project which the client considers unsatisfactory on completion (Cherns & Bryant 1984). The reason for dissatisfaction will stem from how the client initially defined the project (often as a result of a power-driven political process within the client organisation) but this will not stop the client believing that it was due to some fault on the part of the project team. Such an outcome can affect the credibility of the project team such that the strength of their reputation is reduced and so may create problems for them on future projects. This effect does not apply only to the project manager but also to individual members of the project team.

An analysis of the construction of a new university showed that the exercise of power within a complex client structure and a sensitive wider political context meant that the project was commenced without a full understanding of the type of university to be constructed (A. Walker 1994; Walker and Newcombe 2000). Whilst the project was completed extremely quickly, the final cost was three times the original estimate. Nevertheless the project was considered to be good value for money. However, the members of the project team were perceived to have created 'a massive cost overrun' and suffered from this reputation resulting in some of them being rejected for future commissions. The 'overrun' was all that was remembered. The final cost would not have been seen as an overrun if the power plays within the political context had not resulted in a commitment to an extremely short period for design and construction and the resulting lack of cost management.

Power can, of course, work for good and bad. Whilst the above illustration shows that the outcome was negative for the consultants, the exercise of power was positive in delivering a new university three years earlier than expected to the benefit of the community as a whole. The potential for the use of power within public client organisations is probably even greater than for private sector projects. The Hong Kong Convention and Exhibition Centre took ten years from conception to the start of design due to competing power plays by government departments. Only when a chairperson held in very high regard in the community and reinforced with charisma was appointed to the Trade Development Council (TDC) was sufficient power vested in the TDC to enable it to break the deadlock (Walker & Kalinowski 1994).

Whilst focusing on project management generally within client organisations, rather than construction in particular, Pinto (2000) considers politically sensitive project managers understand that the disruption caused by the development of a new project reshuffles the distribution of power within an organisation. Thus project managers need to be aware of the forces at play. He believes that in order for their projects to be implemented successfully 'most managers are forced through expedience to adopt methods for influence and politics'. However he does not define what he means by a successful project. The implication seems to be completing a project that is accepted within the power structure of the organisation. This illustrates the difficulties faced by project managers external to the client organisation and by project teams in defining and accomplishing projects.

A distinctive power source on construction projects is that of patronage. This stems from the process of obtaining commissions for future projects. Whilst increasingly commissions are awarded on the basis of competitive bids, particularly in the public sector, nevertheless in the private sector commissions are frequently awarded on the basis of recommendations. Even for competitive bids, recommendations are often needed in order to be admitted to the list of bidders. Traditionally architects had the gift of selecting the consultants with whom they would work. In contemporary times, this gift has in many cases moved to the project manager. This situation gives the project manager or the architect enormous power over the consultants as they will be unwilling to go against the desires of those who may hold the gift of future work. Even on projects where consultants have been appointed in competition the position is no different as the process may introduce consultants to project managers or architects with whom they have not previously worked. The potential of future commissions from new sources will act to give power to the project manager or architect.

The sources of power described earlier (reward, coercive, expert, referent) need relating to construction. In the management literature reward and coercive power are seen to be legitimate power (or authority) as they offer benefits or sanctions which operate within the formal structure. However it has been argued above that this may not be the case for construction projects. Rewards and coercion can be applied by those outside the formal authority structure of the project and may thus distort the advice and actions taken by the dependent party where such advice and action is opposed to the view of the power holder yet is in the best interests of achieving the project's objectives. Project managers need to be very sensitive to such issues.

Expert power is a major force on construction projects. Projects are so technologically and managerially demanding that reputation and track record count for a great deal. The specialisation of professional skills also contributes to the effectiveness of expertise as a power base. Project managers and others have great difficulty in contradicting the expert advice of a contributor even though they may instinctively believe that it is inappropriate. A corollary is that such experts are likely to gather support from colleagues against less expert members of the team who may disagree with the expert. Liu *et al.* (2003) make the point that the project manager requires sufficient power to question inputs from specialists, if only from an integrative perspective, given that a majority of problems are due to missing, incorrect or unco-ordinated information.

Referent (or prestige) power is the influence that people exercise because people believe in them. Although closely related to expertise power it is more associated with a record of substantial achievement. This source of power is particularly reinforced by charisma. It is a particularly potent force on projects at a number of levels. At the initiation of projects people with this type of power are able to force through the acceptance of projects 'against the odds'. For example the selection of one airport site as opposed to another. Project managers with referent power are naturally in a strong position as are other members of the project team, for example an architect with an international reputation for the design of famous buildings. Project managers and others who worked on the Mass Transit Railway in Hong Kong, which was an extremely successful project, gained great referent power and went on to obtain other important project management and other posts as a result.

At another level, a play which relies on referent power is when one of the consultancy firms sends along a director to a project meeting which would normally be attended by a project level staff member, for the purpose of obtaining a particular outcome. The intimidation of other members of the meeting by the presence of a director can be seen as the use of referent power.

Architects as a profession gain referent power from their historical social position. This certainly still occurs in the UK, though probably less than in the past, and no doubt in other countries. In the early part of this century architects emerged as the elite social class associated with construction stemming from the patronage of their clients as described by Bowley (1966). Architects continue to derive referent power from this historical context.

Finkelstein (1992) refers, in passing, to prestige power also being provided through a manager having powerful friends. This facet assumes a more significant position in construction. Due to the high interdependence of the elements of the process both inside and outside the project management system, the use of 'powerful friends', or rather better put as 'contacts with influence', plays a large part in moving the project forward at all stages. Such influence can include, for example, arranging finance at the early stages, contacts with government to ease the approvals process and good contacts with suppliers and subcontractors during construction. People who can get things done through their contacts acquire strong prestige or referent power.

A clouding of the distinction between authority and power on a conventionally organised project is featured in Loosemore's (1999) paper in which it is stated that, '...expert and legitimate contractual power plays a minor role...'. The reference here is to the authority given under a standard form of contract and not the real meaning of power (when it is used outside the legitimate, or formal, authority structure). Nevertheless, the case study presented is interesting as it shows the way in which responsibility under the contract is allocated bears little resemblance to expert and legitimate contractual authority but more to the way in which power struggles between interest groups resolve themselves.

Politics, projects and firms

Previously construction project organisations were classified (using Mintzberg's taxonomy) as shaky alliances of professional and innovative organisations, the shaky alliance being one of Mintzberg's types of political organisation.

It is worth continuing this discussion at this point. Firstly, this configuration applies to projects for which the project team is not wholly in-house. If the project team was wholly in-house it could be part of a political organisation which fitted any of the other three forms – confrontational, politicised or complete political arena. In such circumstances the development of an effective project could be difficult to achieve although perhaps less so in a politicised organisation. A project team not wholly in-house would be unlikely to fit such descriptions.

However, even when not wholly in-house, these typologies have relevance for understanding the effect of organisational politics on projects. The client organisation could be of any of these forms at the time the project is being initiated and objectives are being defined. The possibility of obtaining a suitable brief must be extremely limited in the case of the confrontational and complete political arena forms but the shaky alliance and politicised organisation should be less difficult if understood by the project manager and project team.

The other important elements which have the potential to become political organisations are the individual consultancy firms which make up the project team. Mintzberg would categorise them as professional organisations which he believes is one of the organisational types in which there is considerable room for political games. If so, they have the propensity to become political organisations of a confrontational, politicised or complete political arena type. If a consultancy firm transformed from a professional organisation to a political organisation during the course of a project, the effect on the project could be severe. The likelihood would be that rather than focusing on the project, the attention of members of the firm allocated to the project would be deflected to the internal problems of the firm. In an activity as closely interdependent as a construction project this could be to the detriment of the project.

Whilst the illustration here uses Mintzberg's description, many other accounts of politics in organisation are also good bases for analysis (e.g. Kakabadse *et al.* 2004; Schermerhorn *et al.* 2004; Robbins 2005). An interesting point is that not all writers believe that organisational politics is a negative force. Some believe it enhances the achievement of organisational goals and survival and is good for career advancement (Schermerhorn *et al.* 2004). The important aspect for project management is to recognise that these forces exist within client organisations and will probably have an impact on project definition and client involvement over which project managers (whether in-house or not) will have no, or very little, influence. Similarly such forces will be present in the consultancy firms that provide team members, and may influence their attitude to the project and their performance. These forces could lead to a politicising of the project team, the management of which will be the responsibility of the project manager.

Empowerment and projects

Empowerment is a relatively new buzz-word in management and is important to many organisations which include powerless and disadvantaged groups. However, in the process of managing construction projects empowerment of the project team occurs naturally due to the nature of the task which requires highly skilled professionals to be self-motivated and accept high performance goals. This they have always done due to their professional training and traditions so the concept is not new although it has not been identified and given the name empowerment.

Generally speaking project managers instinctively empower their project teams although little effort is needed as the professional contributors generally empower themselves. Nevertheless this may not occur on every occasion so the concept should be understood and applied by project managers to further enhance the project team's contribution.

However, if the construction team is included in this scenario, Newcombe (1994) claims that the construction management procurement system produces a different power configuration than the traditional procurement system. He believes this system allows a better opportunity for the exercise of expertise power and charisma coupled with the empowerment of all the parties especially the specialist trades contractors. Newcombe (1996) extended these views by contrasting what he saw as the hierarchical traditional system of project organisation with the 'potentially democratic' construction management form. He believes the latter is based on empowerment or power equalisation and that such power structures are to the benefit of clients, as they will not produce the fragmentation and friction created by the traditional system. He argues:

> 'If the empowerment approach is adopted then skill in building networks of contacts such as designers, trades contractors, clients, suppliers and stakeholders with an interest in the project will be necessary'.

These ideas reflect the current preoccupation of the construction industry with various forms of partnering but empirical work is required to substantiate them. Illustrating the unevenness of empowerment in traditional organisation structures, Hammuda and Dulaimi (1997) state that construction managers were always 'traditionally empowered' on their sites by making decisions in relation to their autonomous project within their wider companies' boundaries but that this is becoming less so as projects become more complex. In comparing construction with service and manufacturing industries, they find that empowerment in the construction industry (by which they refer to companies) is very much behind.

Williams (1997) tempers enthusiasm for empowerment by pointing to the conflict between it and risk management. He believes that 'unfettered empowerment in a modern project setting . . . can lead to high risks'. He continues:

> '. . . some Risk Managers, in projects which have embraced the management philosophies of both empowerment and risk management, have found themselves in an almost schizophrenic role, required to manage risks others are

empowered to effect or enhance. What are needed, are compromises, empowering project workers and teams whilst recognising the holistic nature of the projects.'

Whilst not referring to construction specifically, nevertheless his message is clear.

Project managers and power

Project managers need to recognise the power configurations in projects. They need to recognise that power has a potential for influence over others outside the legitimate authority structure. If such influences are seen to be harmful to achieving project objectives, the project manager needs to work out strategies to deal with them. At the same time the project manager needs to avoid becoming paranoid over power which is easily done when working in complex organisations. Power plays, if they exist, need turning to positive not negative ends.

Pfeffer (1992) has constructed some questions to be used by managers for understanding the pattern of power in organisations which are also useful for project managers:

- 'Decide what your goals are, what you are trying to accomplish.
- Diagnose patterns of dependence and interdependence; what individuals are influential and important in you achieving your goal?
- What are their points of view likely to be? How will they feel about what you are trying to do?
- What are their power bases? Which of them is more influential in the decision?
- What are your bases of power and influence? What bases of influence can you develop to gain more control over the situation?
- Which of the various strategies and tactics for exercising power seem most appropriate and are likely to be effective, given the situation you confront?
- Based on the above, choose a course of action to get something done.'

Chapter 9
Project Leadership

Introduction

In the first chapter care was taken to distinguish between the title 'project manager' and the process of 'project management'. Subsequently focus has been principally upon the process with relatively infrequent references to the role of the project manager. The last two chapters have looked at the activities of project management and authority in projects so that it is now appropriate to consider certain aspects relevant to the person who is charged with making the organisation work effectively. For convenience this person will be referred to as the project manager, but may go under a number of titles in practice, and will refer to the person managing the project for the client.

This chapter will not set itself the impossible task of describing the perfect project manager – that holy grail that all should seek. Rather it will examine the major aspect of performance of project managers which aids effective management – leadership.

Leadership is an intrinsic part of management. It is the manner in which managers conduct themselves in their role in order to obtain the best performance from the people they are managing. Hence all managers lead, some much more effectively than others. Leadership takes many forms depending on the characteristics of the leader and the demands of the situation in which it is exercised. It is rarely needed in the form of the public perception of it – the 'knight on a white charger' – but much more frequently as perceived by the Chinese philosopher Lao-tzu – 'to lead the people, walk behind them'.

Interestingly, after hundreds of years, Lao-tzu's philosophy seems to be reflected in recent work on leadership by the recognition of 'servant leaders' (Howell 2006), whose primary concern is for their followers and who 'place service before self-interest, and are not concerned with acquiring power, prestige, financial reward or status'.

An understanding of the way in which the actual application of leadership skills to project teams takes place is necessary for project managers and this is examined. The distinctive demands of managing projects as opposed to managing in general require an interpretation of the literature and practice of general leadership to render it useful to project managers.

Leadership

Much has been written and a number of major research studies have been undertaken on leadership, without a convincing outcome (McCabe *et al.* 1998). It is generally accepted that leadership is important for management to be effective and this tends to create a general perception that management and leadership are somehow separate. This point has been made by Fewings (2005), but in reality the distinction is not usually apparent. Management is seen to be a mechanical process using techniques, responding to directives from elsewhere and controlling those being managed. Leadership is perceived as charismatic, inspirational and forward thinking. However, the definition of management used in this book includes the phrase 'working through others'. Immediately this is recognised, leadership and management become intrinsically linked.

The public perception of management, as a mechanistic process, is better described as administration. This was well stated many years ago by Townsend (1984) and is still relevant today. He believes that 'most people in big companies today are administered not led. They are treated as personnel, not people' and he quotes *The Peter Principle* (Peter & Hull 1969) in support, 'most hierarchies are nowadays so cumbered with rules and traditions and so bound in by public laws, that even high employees do not have to lead anyone anywhere, in the sense of pointing out the direction and setting the pace. They simply follow precedents, obey regulations, and move at the head of the crowd. Such employees lead only in the sense that the carved wooden figurehead leads the ship.' Leavitt (2005) believes that this continues to be the case and says: 'Leadership has become the most widely discussed, widely taught, and widely written about subject in the world of organisational studies. It may also be the foggiest'.

Leader and manager cannot be separated. The quality of a manager's leadership will be demonstrated by the quality of achievement of the people he or she leads. The dangers of having a project administrator rather than a project manager or project leader are all too apparent from the foregoing. Project managers should lead not administer project teams if they are to be effective in delivering projects which are functionally effective, on time and within budget. But this does not have to be high profile, charismatic leadership, it may be much more effective in another of Lao-tzu's conceptions, 'As for the best leaders, the people do not notice their existence. The next best, the people honour and praise. The next, the people fear; and the next the people hate... When the best leader's work is done the people say "we did it ourselves".'

Underpinning leadership is the authority and power profile of the leader (as discussed in the last chapter and not repeated here), which plays a large part in the leader's effectiveness. Formal authority, reward and coercive power are likely to be greater forces in leading non-project organisations, whereas a leader's formal authority reinforced by expert and referent power are more significant in leading construction project organisations (Liu *et al.* 2003).

A further aspect discussed previously that has an impact on leadership effectiveness is trust; that is, trust in leaders. Dirks and Skarlicki (2004) find that 'surprisingly little research has focused on illuminating *how* trust in leaders contributes to the effective functioning of groups and organisations and how it can be leveraged to meet this objective'. They continue by saying that the evidence on the issue is mixed and that many ideas await empirical testing. Robbins (2005) believes that 'trust appears to be a primary attribute associated with leadership'. He identifies three types of trust: deterrence-based, knowledge-based and identification-based, which correspond closely to the power categories of reward (and coercive), expert and referent.

In what management gurus may see as heresy, the opinion has been expressed that leadership may not always be important (Robbins 2005) and even that leaders' actions may be irrelevant. Examples of such situations are when tasks are unambiguous, when goals and rules are formalised and when those being 'led' are experienced professionals (with implications for construction projects). In a similar vein, Schermerhorn *et al.* (2004) quote Pfeffer (1981), who contends that leaders of large corporations have little impact compared with environmental forces and that much of the impact of a 'top leader' is symbolic. Similarly, symbolic treatment of leadership occurs when many people could have been responsible for performance. Schermerhorn *et al.* quote Meindl's (1990) naming of this phenomenon 'the romance of leadership, whereby people attribute romantic, almost magical, qualities to leadership'.

Some research models

It used to be believed that leaders were born not made, a myth no doubt encouraged by the leaders themselves. Whilst research has shown that leaders are not born as such, it has not been able to show clearly how leaders do emerge. Certainly personal characteristics do contribute to the ability of people to lead but personality develops through experience and is only partly hereditary.

Nor is there agreement on the essential qualities of a leader. This is so even in the military field, which has a well developed sense of leadership by its definition. The following lists of qualities of leadership from four military establishments illustrate the point:

Navy	Air Force	Army	Marines
Faith	Efficiency	Bearing	Integrity
Courage	Energy	Courage	Knowledge
Loyalty	Sympathy	(Physical and moral)	Courage
Sense of duty	Resolution	Decisiveness	Decisiveness
Integrity	Courage	Endurance	Dependability
Humanity	Tenacity	Initiative	Initiative
Commonsense	Personality	Integrity	Tact
Good judgement		Judgement	Justice
Tenacity		Justice	Enthusiasm
Fortitude		Loyalty	Bearing
Physical fitness		Tact	Endurance
Mental fitness		Unselfishness	Unselfishness
Self-control			Loyalty
Cheerfulness			Judgement
Knowledge			

If the military have such variety, it can be seen that the difficulties facing researchers into leadership in business and industry are enormous.

As mentioned earlier, research has not been able to show clearly how leaders do emerge, nevertheless, some valuable directions and tendencies have been identified which are useful to managers who are seeking to improve their leadership skills.

Much leadership research stemmed from the work of Fielder (1967) in the late 1960s. His model argues for a contingency orientated approach in which two factors were particularly important:

(a) the degree to which the leader is liked and trusted by the group
(b) the degree to which the group's work is defined.

The contingency approach is best illustrated by polarising two styles of leader:

(a) the directive managing, task orientated leader
(b) the non-directive human relations orientated leader.

Fielder argued that each style was appropriate in particular situations. In very favourable situations, i.e. when the leader is well liked and the group's work is clearly defined, and in very unfavourable situations, i.e. the opposite of the last, a directive management approach is better. Conversely if the leader was well liked but the group's work unstructured and when the leader is not liked but the group's work is clearly defined then a non-directive approach would be beneficial. This is a very severe simplification but gives a sense of the developments in thinking about leadership which were engendered by Fielder.

This work had been preceded by Likert's which concluded that the best supervisors were employee centred. The range of styles implied by Fielder's work placed Likert's into perspective by recognising a directive management approach in certain circumstances.

The development of the idea of the usefulness of a range of leadership styles, each to be used in the appropriate setting, was demonstrated by Tannenbaum and Schmidt (1973). A simplified range is given in Fig. 9.1.

This work implies that the leader needs great sensitivity to the situation regarding the feeling of the group being led and the nature of the task being undertaken and needs to be sufficiently flexible to change styles as appropriate.

This approach is further illustrated by the work of Hersey and Blanchard (1972) and of Blake and Mouton (1978) who considered leadership styles as an amalgam of task and relationship orientated approaches (equivalent to the directive and non-directive styles referred to earlier). The appropriate mixture to be used should be determined by the maturity of the group being led and the needs of the situation. Maturity is defined in terms of achievement, motivation, the willingness and ability to take responsibility, education and experience. It is suggested that leadership behaviour should relate to high/low task and high/low relationship permutations as illustrated in Table 9.1. The relevance of these permutations to the maturity of the group being led is illustrated in Fig. 9.2.

Bonoma and Slevin (1978) have used information input and decision authority in formulating their graphical model of leadership. They believe that the leader's

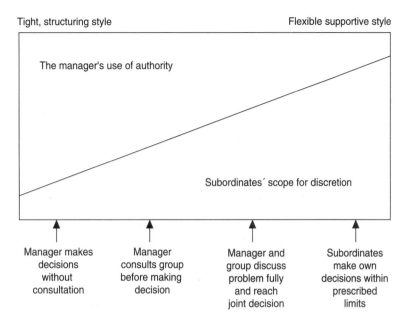

Fig. 9.1 Some styles of leadership (adapted from Tannenbaum & Schmidt 1973 by Fryer 1985).

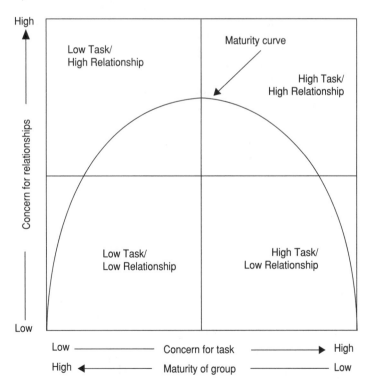

Fig. 9.2 Effective leadership styles versus group maturity. (*Sources:* Hersey & Blanchard 1972; Blake & Mouton 1978; Gilbert 1983.)

Table 9.1 Basic leader behaviour styles. (*Sources:* Hersey & Blanchard 1972; Blake & Mouton 1978; Gilbert 1983).

Basic styles	Situation	Effective leadership	Ineffective leadership
High task and low relationships	Efficiency in operations as a result of arranging conditions of work in such a way that human elements interfere to a minimum degree	Seen as knowing what he wants and imposing his methods for accomplishing this without creating resentment	Seen as having no confidence in others, unpleasant and interested only in short-term output
High task and high relationships	Work accomplishment is from committed people, whose interdependence, through a 'common stake' in organisation purpose leads to relationships of trust and respect	Seen as satisfying the needs of the group for setting goals and organising work but also providing high levels of socio-emotional support	Seen as initiating more structure than is needed by the group and spends more time on socio-emotional support than necessary
High relationships and low task	Thoughtful attention to the needs of people for satisfying relationships leads to a comfortable friendly organisation atmosphere and work tempo	Seen as having implicit trust in people and as being primarily concerned with developing their talents	Seen as primarily interested in harmony and being seen as 'a good person' and unwilling to risk disruption of a relationship to accomplish a task
Low task and low relationships	Exertion of the minimum leadership effort to accomplish the required work is appropriate to sustain organisation membership	Seen as appropriately permitting his subordinates to decide how the work should be done and playing only a minor part in their social interaction	Seen as uninvolved and passive, as a 'paper shuffler', who cares little about the task at hand or the people involved

need for information from the group he is leading and also where the leader allows the decision authority to lie, are the factors which determine the leadership style to be adopted. Their graphical model is illustrated in Fig. 9.3.

As Slevin (1983) states:

'Using this plotting system, we can describe almost any leadership style. However, the four extremes of leaders you have known (depicted in the four corners of the grid) are the following:

(1) Autocrat (10, 0). Such managers solicit little or no information input from their group and make the managerial decision solely by themselves.

Fig. 9.3   Bonoma–Slevin leadership model.

(2) Consultative Autocrat (10, 10). In this managerial style intensive information input is elicited from the members, but such formal leaders keep all
substantive decision-making authority to themselves.

(3) Consensus Manager (0, 10). Purely consensual managers throw open the
problem to the group for discussion (information input) and simultaneously allow or encourage the entire group to make the relevant decision.

(4) Shareholder Manager (0, 0). This position is literally poor management.
Little or no information input and exchange takes place within the group
context, while the group itself is provided ultimate authority for the final
decision.'

Given the four positions in the Bonoma–Slevin model, the leaders should be
able to move their style around the graph in response to characteristics of themselves and the situation they face. Bonoma and Slevin (1978) have tabulated the
types of pressures that leaders might face and the directions that these pressures
might push them. Using the terminology North, South, East and West, you can

Table 9.2   Three leadership style pressures. (*Source:* Bonoma, Thomas V. and Slevin, Dennis P. *Executive Survival Manual*, Wadsworth Publishing Company, Belmont, CA, 1978, p. 89.)

1. PROBLEM ATTRIBUTE PRESSURES	DIRECTION OF PRESSURE ON LEADERSHIP GRID
• Leader lacks relevant information; problem is ambiguous.	North – more information needed.
• Leader lacks enough time to make decision adequately.	South and east – consensus and information collection take time.
• Decision is important or critical to organization.	North – information search maximized.
• Decision is personally important to leader.	North and east – personal control and information maximized.
• Problem is structured or routine.	South and east – little time as possible spent on decision.
• Decision implementation by subordinates is critical to success.	West and north – input and consensus required.

2. LEADER PERSONALITY PRESSURES	
• Leader has high need for power.	East – personal control maximized.
• Leader is high in need for affiliation or is 'people oriented'.	North and west – contact with people maximized.
• Leader is highly intelligent	East – personal competence demonstrated.
• Leader has high need for achievement.	East – personal contribution maximized.

3. ORGANIZATION/GROUP PRESSURES	
• Conflict is likely to result from the decision.	North and west – 'participative aspects' of decision making maximized.
• Good leader–group relations exist.	North and west – group contact maximized.
• Centrality, formalization of organization is high.	South and east – organization style matched.

look at the graph and follow the direction in which each pressure might tend to move the leader (see Table 9.2).

More recent work has linked leadership style to organisation culture (Quinn 1988) in that leaders should function in a manner consistent with the desired culture of the organisation. However, whilst there exist many definitions of organisation culture, there is not yet an accepted approach to research into organisation culture so its relationship to leadership styles is not yet established (Maloney & Federle 1991).

International projects place extra demands on leadership. The Rowlinson *et al.* (1993) examination of leadership styles in mixed culture (in the ethnic sense) construction projects showed that project leaders in Hong Kong have more concern with maintaining good relationships and a harmonious working atmosphere than in the West.

The key to successful leadership is knowing what your dominant style is and being able to modify that style depending upon the contingencies of the various leadership situations that you face, which are likely to include your organisation's culture.

Leadership style

One thing in common between the researchers into leadership is the recognition of different management styles along a spectrum from authoritarian to democratic, although different terms are sometimes used, for example 'task centred' or 'employee or people (relationship) centred'. At the extreme ends of the spectrum the authoritarian manager issues orders without consultation and the democratic manager allows the group to take the decision after having fully discussed the issues. Much of the research suggests that the situation to be managed should determine at which point on the spectrum the leader should be.

The practical application of these ideas does create problems. They presuppose that the leader's leader has the ability to determine the type of leader needed in each situation and to select such a person. This also implies that either if the situation changes the leader should be changed or the leader should change style if the situation changes. The latter seems to be favoured by the writers but it is extremely difficult for an individual to change management styles and for the group, assuming it does not change, to accept different behaviour styles from the same leader. By the time someone reaches a high management level, their character and personality will already have been established and the manager is likely to have settled in a position on the autocratic/democratic spectrum and have a relatively short span of flexibility. If a wide range of flexibility is to be expected, this has major implications for the way in which both project managers and project teams are trained for the construction industry as this degree of flexibility does not exist at present.

Transactional and transformational leadership

The research models just discussed are the basis of the transactional leadership school, which involves interaction between leaders and subordinates to achieve better performance by paying attention to details and being both firm and flexible as appropriate. Leaders guide their followers in the direction of established goals by clarifying role and task requirements (Robbins 2005). Transactional leadership includes the trait approach, the behaviour approach, contingency model and situational leadership.

The transformational leadership school sees itself going beyond the transactional school to inspire their followers to encompass change and elevate their interests beyond their own 'narrow concerns'. The essential characteristic of transformational leaders is charisma. Kakabadse *et al.* (2004) see the concept of charisma as including 'inspiration, adornment and an unequal relationship between the leader and followers', and believes that 'some writers see charisma as attribution, namely the qualities given to a leader by the followers (sometimes because of their own inadequacies). Others consider charisma as emanating from the individual, innate qualities that people find attractive'. Howell (2006) sees this type of leader as 'typically an excellent communicator, giving inspirational speeches and describing highly desirable visions of the future'. Leaders who are frequently identified as this kind tend to be politicians rather than business leaders, although Steve Jobs (co-founder of Apple Computer) is often cited. Within the same school, Kakabadse *et al.* also include narcissism as the opposite to charisma, describing this as 'the tendency of some successful leaders to indulge in vanity and selfishness'.

However, the hype surrounding charismatic leadership overstates its general value. It is patently not generalisable as relatively few organisations would benefit from such leadership, which is valuable only in very specific circumstances; hence it can be seen to be situational. Robbins provides a telling critique:

> 'Even charismatic leadership may not be the panacea that many in the media and public think it is. Charismatic leaders may be ideal for pulling a group or organisation through a crisis, but they often perform poorly after the crisis subsides and ordinary conditions return. The forceful, confident behaviour that was needed during the crisis now becomes a liability. Charismatic managers are often self-possessed, autocratic, and given to thinking that their opinions have a greater degree of certainty than they merit. These behaviours then tend to drive good people away and can lead their organisations down dangerous paths.'

The relevance of charismatic leadership to construction project management can be seen to be similar to its application generally. That is, there may be situations when a crisis which requires charismatic leadership occurs on a project but these are unlikely to be frequent and tend to be restricted to high-profile prestige projects. Even in such an event, it is a long way from certain that a charismatic project manager capable of resolving the crisis could be found.

A comparison of the effect of transformational leadership on project managers and line managers found that this management style had less impact on project managers than on line managers (Keegan & Den Hartog 2004). It is suggested that this is due to the way that this style of leadership is received by those being led, and that the 'career support system supports the line managers in their efforts to motivate and win support'. But this seems to reflect transactional rather than transformational leadership styles. It is also worth noting that all participants in this study were in-house to a government department; construction project team members from various firms are perhaps even less likely to respond to transformational leadership.

Leadership and the project manager

As discussed, there is no definitive research on leadership. What the research has shown is that there are a large number of variables which contribute to determining the style of leadership needed in a given situation in order to manage effectively. They include whether the leader is liked, how closely the work of the group is defined, the maturity of the group being managed, the location of decision authority and the degree of information input. The research reviewed is only a sample of the research which has been undertaken on leadership, hence the variables listed are by no means exhaustive. They are used to illustrate the sort of features that managers should consider when thinking about the effectiveness of their leadership.

Most research on leadership in organisations has focused on general management or line management situations rather than on project management. It has already been illustrated that the management of a construction project has unique features such as the relationship with the client and the interorganisational nature of the process. The project manager is usually in the position of leading contributors from a large number of commercially independent organisations over whom he has only limited formal authority. A significant outcome of this situation is that, in the matrix, each contributor will be subject to 'leadership' from both the project manager and the manager of the employer's organisation. In effect the 'leaders' are in 'competition'. As long as the leaders have the same objectives there should be no problems but the potential for the contributor's leader from the employing organisation to distort the project manager's leadership is present.

Project managers will be leading a group of mature, experienced professionals and are often only slightly elevated over them in terms of legitimate authority. They may even resent having project managers in this position. Consequently the project manager's leadership style will tend to be democratic and rely on influence and persuasion rather than formal authority. Project managers are continually in receipt of information from their teams and must interact with them, constantly exchanging large volumes of information of a technical and financial nature. A not dissimilar relationship exists with the client with the important exception that the client has final formal authority over the project manager. Nevertheless a similar style, based on persuasion, will usually be appropriate for the project manager, although it can be seen as a 'subordinate' leading a 'superior'.

However, empowerment of project teams by a project manager and of a project manager by a client can produce contrasting reactions from project leaders and project team members (Liu *et al.* 2003). Leaders may feel threatened by empowered subordinates due to loss of control, and not all subordinates want to be empowered as they may not wish to accept the responsibility that goes with it.

Each of the subsystems in the construction process terminates in either a key or operational decision. The nature of the work to be undertaken within each subsystem will determine the leadership style most appropriate to arriving at the best outcome. However in each case the decision will effectively be made by the project manager rather than the project team, although in the case of key decisions it will be in the form of a recommendation to the client. So whilst for some subsystems the project manager should adopt a democratic form of leadership as ideas are formed and developed, at the end of the process it will be the project manager acting 'autocratically' who will make or recommend a decision. It will be the project manager's responsibility to do so although in practice it may not appear to be autocratic as in many cases the decision will be arrived at by consensus. However, many project managers may not wish to appear in such a high profile decision-making role and may seek to have decisions seen as group decisions. There are degrees of democracy and the actual position on the scale taken by project managers will depend on a number of factors such as the urgency of the decision, their personal characteristics and the experience of the group being led. The subsystems which lend themselves to a democratic style of leadership are those concerned with developing initial evaluations, outline design strategies and selecting contractual arrangements.

Other subsystems may lend themselves to a leadership style which is more towards the autocratic end of the scale. These will include developing working drawings, preparing contract documentation, cost control during construction and many during the construction stage if let by conventional competitive tender. However, should problems occur on any such mechanistic subsystems it may be that a more democratic style of leadership may be necessary to enable the group to solve the problem. Conversely within a subsystem being led democratically it may be necessary on occasions to be autocratic to bring it to a conclusion and make progress.

The project manager has the taxing job of playing the scales from democratic to autocratic leadership depending on the needs of the particular subsystem or the particular stage of development of the subsystem. Anderson (1992) confirmed this view and found that leadership skills had the highest frequency of significance across eight project management functions examined yet project managers were perceived as having only average or less than average managerial skills. However, it should be recognised that the project manager's need for flexibility is extremely high not only because of the nature of the process being managed but also because of the maturity and organisational independence of the contributors who originate from separate firms. In such a context the project manager's position on the scale will lay predominantly towards the democratic end as he or she seeks to cope with what is inevitably a challenging leadership position.

Results of a study of the power base of project managers of chemical installations in China (Liu & Fang 2006) reflected these ideas, but at a much finer level of analysis. They found that good project leadership operates through managing the gaps between the amount of power in a position in the project team and the power actually needed to get the job done through power-sharing/empowerment and power-amassing. Importantly, they found that:

'. . . it is not positional power [formal authority] . . . but the manager's charisma (referent power) and the discrepancy of expert power between the project manager and team members that instigate the manager's behavioural flexibility in managing the power gap. Such characteristics relate, presumably, to the nature of project organisations, for example, multi-disciplinary, temporary and inherent with uncertainties and interdependencies.

The positional power of project managers plays a relatively weak part in the model.'

They suggest that this may be due to the only marginal additional formal authority a project manager has over team members and that 'project team managers in China are delegated little positional power in the existing social system . . . '.

Whilst the demands of project management may be rather different from general management, it will be helpful to project managers for them to review the previously described research models. This will enable them to ask themselves questions about how they lead project teams, and help them to focus on the variables which have been seen to have an effect on leadership.

Fielder's (1967) work makes a good beginning by asking project managers to consider how well liked they are by their teams and also to evaluate the degree of uncertainty of the task being undertaken. The latter is very variable in construction, ranging often from practically 'a blank sheet of paper' in the early stages of some projects to what can be a well defined task of constructing the building. In contrast for some standard building types the degree of certainty of the tasks is very high. It also asks project managers to assess where they see themselves on a scale from autocratic to democratic manager and the degree of flexibility which they consider themselves to have. Unfortunately these are not questions which we can easily answer about ourselves nor are they opinions we like to hear from others. Herein lies one of the greatest dilemmas in improving leadership abilities – that of recognising honestly one's own abilities.

The flexibility of leadership styles needed by project managers is clearly demonstrated by Tannenbaum and Schmidt's scale (Fig. 9.1). It is clear that the wide range of situations encountered on construction projects will require styles covering much of the scale, although there will be a strong bias towards the right hand end due to the appropriateness of professionals making decisions within their competence. This begins to illustrate that project managers will need to have a fundamentally democratic style of management.

The low task and low relationship leadership style of Blake and Mouton is clearly appropriate to many construction projects for which the project manager is dealing with highly educated, well trained and emotionally mature people. It also demonstrates other modes which may be necessary in specific circumstances, particularly if a crisis situation occurs. However, such situations are rare and the project manager is more likely to have to switch to a low task/high relationship, or even a high task/high relationship mode, if leading a somewhat less mature group particularly in times of high construction activity when less experienced staff occupy more senior positions.

The range of leadership styles is well illustrated in Bonoma and Slevin's model (Fig. 9.3) and is particularly useful for project managers who are dependent upon their team of professionals to provide them with information and who frequently use discussion with the team as a means of arriving at a decision. Construction project managers will mainly operate along the top of the grid between the 'consensus manager' and the 'consensus autocrat' but may at certain times find themselves having to adopt styles located at other points in the grid.

The relationship between leadership and organisation culture has been referred to. The problem with this concept in relation to project organisations is that analysis of the organisational culture of construction project organisations does not appear yet to have been undertaken so its impact on leadership is not known. The concept of organisation culture, in relation to temporary organisation structures rather than permanent evolving business organisations, is likely to need special consideration as no sooner will a culture begin to grow than it will be dissipated. Its effect on management style may be limited although the development of even a limited culture is far more likely to be affected by the management style of the project manager.

The varying requirements of leadership styles are readily apparent in project management organisations. For example:

- At the early stages of a project, the manager has to weld together the range of professional specialists involved in the project and lead them in balancing the conflicting objectives which will no doubt have emerged from the client organisation. Developing a viable solution which balances function, cost and time requires the project manager to create a climate in which everyone is free to speak up, make suggestions and criticise. To achieve this the project manager must adopt a human relations orientated style of leadership. Once the proposed solution has been defined, the process of developing the solution into working documents becomes a much more structured process requiring a more task-orientated leadership style.

- During the process of developing working drawings, contract documentation, etc., which requires a task orientated approach, it may become apparent that a major problem occurs in translating the proposed solution, in which case, the project manager is unlikely to be able to solve the problem alone. The project manager will need to bring together professional specialists to discuss and resolve the problem and the project manager's style will switch from a task to human orientated style until the routine work is recommenced.

- The project manager's formal authority affects the style adopted. During the construction stage of a project under traditional contract conditions, the project manager will normally have to adopt a directive or autocratic style as the contractual context places specific obligations upon him or her. Such conditions make it difficult to adopt a human relations style even when it appears to be the best way of approaching a problem. 'Partnering' is an approach which seeks to allow a more human relations style which could be particularly effective in the construction stage.

- The degree of authority can also affect the project manager's style during the pre-contract stage. The authority which is held by the project manager who is an employee of the client organisation or who is given clear authority by the client, provides the project manager with a greater opportunity to be autocratic which is not the case if the project manager's authority is ambiguous. The former gives the project manager a greater range of leadership styles to use if he or she is capable of exercising such flexibility and if he or she has the ability to resist falling back on his or her formal authority in inappropriate situations.

Culp (2004) has an interesting perspective on project leadership. He sees the basic needs of project team members as a 'triangle of needs' formed by a triangle of equal sides which represent content needs, procedural needs and relationship needs. Content needs include scope, budget, expenditure, etc.; procedural needs include monitoring and reporting progress, payments and changes in scope; and relationship needs are about perceptions of trust, commitment, communication, fairness, etc. He sees both management and leadership as essential, but believes someone who only manages is essentially concerned with content and procedure and neglects relationship needs until something goes wrong. That most project failures occur due to relationship issues is not recognised by project managers, but is recognised by project leaders who focus on relationships from the start of the project. His emphasis is on spending an equal effort on each of the three elements, but although each element will need constant attention it is probable that the intensity of attention will need to vary between the elements depending on the particular stage of the project.

Project managers' perceptions

It is really not possible to separate project managers' instinctive leadership styles from their view of the people they are leading. As a result the degree of flexibility of leadership styles of which a manager is capable may be inhibited by the fundamental assumptions made about the people being led.

The early work in this area was undertaken by Maslow (1954) who developed the famous 'hierarchy of people's needs' as illustrated in Fig. 9.4. This work defined what people were seeking from work and hence what motivates them and identified that:

- The needs hierarchy is based on needs, not wants.
- It operates on an ascending scale. As one set of needs becomes fulfilled, the next higher set comes into play.
- We can revert to a lower level. For instance, a person operating at level 4 or 5 will fall to level 2 if a feeling of insecurity takes over. Once the need is met, however, he will return to his former needs area.
- Needs not being met are demonstrated in behaviour. To create an environment in which motivation can take place, managers must therefore be able to recognise behaviour patterns in individuals and work groups. This means developing the ability to 'read' people and situations.
- To avoid apathy, which finally results when needs are unfulfilled, managers must be able to implement the right action at the right time.

This fundamental work was developed further by McGregor (1960) and resulted in his Theory X – Theory Y model referred to previously in Chapter 3. He discovered that managers' thinking and approaches were based on two different sets of assumptions about people. A Theory X manager instinctively adopts an autocratic leadership style because he assumes, amongst other negative features, that people are lazy and wish to avoid responsibility. A Theory Y manager instinctively adopts a democratic style as he assumes, amongst other positive features, that people are self-motivated and wish to achieve and enjoy responsibility.

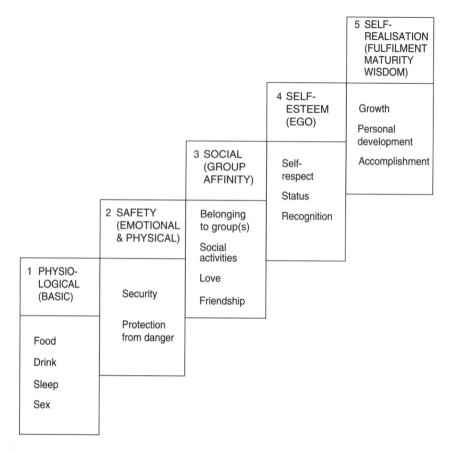

Fig. 9.4 Maslow's hierarchy of individual needs (Maslow 1954).

Herzberg's (1968) subsequent work on motivation reflected the same perceptions and he identified that by removing factors causing dissatisfaction a manager
removes unhappiness but does not generate motivation whereas he was able to
identify the factors which motivated people which correspond to Maslow's levels 4 and 5.

Each project manager will have a particular view of the people being led which
will be somewhere in the Theory X – Theory Y scale and which will determine
the project manager's habitual or instinctive leadership style. Recognition of this
by project managers may allow them to be more flexible but if they are faced
with situations in which they are uncertain they are likely to fall back upon their
instinctive leadership style.

It is important to recognise that whilst the ideas of people's attitude to work
are intellectually satisfying and difficult to disagree with at that level, they are
difficult to apply consistently in practice. Eilon (1979) has pointed out that, if
everyone is self-actualising then the task to be achieved is likely to be neglected
when it is routine but vital to success.

Whilst the argument for leaders to use flexible leadership styles may be acceptable, there is an equally strong argument for those being led to adopt flexibility in their expectations from their work and from their leaders.

This work, whilst still useful, was the forerunner to the broader field of organisation behaviour. It should be pointed out that it is generally held that there is no such study as behaviour of an organisation but only a study of the behaviour of individuals in organisations (Naylor *et al.* 1980). The literature in the field draws heavily on industrial/organisational psychology and the social sciences, one of the more important aspects of which is motivation. Much of the focus is on the behaviour–performance–outcome relationship (e.g. Vroom 1964; Porter & Lawler 1968; Campbell *et al.* 1970; Naylor *et al.* 1980) which believes that the attractiveness of outcomes and individuals' expectation of success determine the amount of effort that they are willing to apply to achieve the goals. They also evaluate their own performances against the expected outcomes. An example of an application of such ideas to construction is given by Liu (1999) who models the effect of project complexity and goal commitment on project outcomes. Such themes are far more complex and sophisticated than those described earlier but it is not possible to develop them further here. However Maslow's ideas continue to provide inspiration for research in construction (Shoura & Singh 1998) even though the field has moved on a considerable way. There is little doubt that they are more easily digested and applied by practising managers.

Leadership qualities

What then are the qualities which a project manager should possess to be a good leader? What are the qualities which allow the project manager to be sensitive to the different situations which arise on a project and be sufficiently flexible to use an appropriate leadership style?

These qualities can be split into characteristics and skills. The project manager's characteristics will in many cases determine how well he employs his skills. Examples of the characteristics which help to form good leaders in construction project management are:

- Integrity
- Preferred leadership style (tending towards democratic)
- Self-confidence
- Ability to delegate and trust others
- Ability to cope with stress
- Decisiveness
- Judgement
- Consistency and stability
- Personal motivation and dedications
- Determination
- Positive thinking
- Excellent health

- Openness and the ability to hear what others say
- Ease in social interactions with many types of people.

In terms of skills, the following are important:

- Persuasive ability
- Negotiation skills
- Commercial expertise
- 'Political' awareness
- Breadth of vision
- Integrative skills
- Ability to set clear objectives
- Communication skills
- Management of meetings
- Early warning antennae
- Skills of diplomacy
- The skills of discriminating important information.

The context of these qualities is the experience of the project manager. A broadly based experience is required of all phases of a project. The project manager will require an appreciation of all the specialist areas, whilst not needing to be a specialist in any. Nevertheless certain aspects will be more significant to the project manager such as contract strategies, cost and time control, money management including project finance and capital and revenue relationships. This type of experience can give project managers the confidence which a leader needs but equally importantly is the way in which it affects the team's perception of its leader. If project managers are seen to have the right experience, or more importantly, they have a successful record of achievement, be it by good leadership or good luck, the team will more readily accept them as leader. Project managers' acceptance in this way can account for much of their perceived leadership qualities.

It is generally accepted that there is no ideal leader for all situations. The variables involved in leadership behaviour and success are wide and unquantifiable. It is only possible to identify them and recognise the kinds of situations in which different combinations of attributes may be most beneficial. It is then necessary to require that leaders have the ability to vary the way in which they amalgamate their attributes depending on the situation in which they find themselves. Failing the leader's ability to be able to do that, it is necessary to change the leader when the situation changes. Neither alternative is really viable although some leaders do have some ability to change their style within limits. A transfer of leadership is for most organisations unrealistic, and which in any case could be extremely disruptive if allowed to happen frequently. The ability to select appropriate leaders in such situations must also be in question.

A respected leader has an amalgamation of characteristics, skills and experience which are recognised and respected by those he leads. Leadership is granted by the people being led and not by an organisational position. The effective leader influences not just subordinates but superiors and peers and in construction project management has the additional need to exercise such influence over people in other companies, which requires leadership characteristics of a very high order.

In trying to answer the question of how one spots a leader it is worth turning to Townsend (1984):

'They come in all ages, shapes and sizes and conditions. Some are poor administrators, some are not overly bright. One clue: since most people *per se* are mediocre, the true leader can be recognised because, somehow or other, his people consistently turn in superior performances.'

Chapter 10
Organisation Structures

Introduction

The traditional approach to project organisation with the architect as team leader, responsible for both design and management of the project with the contractor appointed on the basis of a competitive tender, is still widely used the UK and many other parts of the world. However, variations on this method continue to emerge, as well as some rather more innovative approaches to project organisation.

This chapter examines the contribution of a number of approaches to the solution of organisational problems against the features of project organisation previously identified. These features were:

(a) The relationship of the project team to the client organisation and the client's influence upon the decision points.
(b) The degree of interdependency of tasks and people generated by the project and the organisation structure.
(c) The degree of differentiation present within the operating system (which ideally should be reduced to a minimum). The level to which it can be reduced will be constrained by the nature of the project.
(d) The level of integration provided by the managing system and the complexity of the managing system itself. Over-elaboration can lead to severe differentiation within the managing system, which should have the capability to match its integrative effort to the degree of differentiation present in the project.

Before moving on to examine various organisational approaches, it is appropriate to place them in context by returning to the contribution of transaction cost economics in explaining the theoretical basis for the choice of organisation structure. Whatever organisational structure is chosen requires the features identified above to be considered. Different organisational structures will generate the need for different levels of project management depending upon the nature of the project, its environment and the organisational structure chosen. Project management is not costless and transaction cost economics can relate the cost of project management to the choice of organisation structure.

Project management theory and transaction cost economics

The integration of project organisation theory and the transaction cost approach arises through project organisation theory defining the type and nature of transaction costs by modelling the process, and transaction cost economics providing a framework for analysing the behaviour of the different parties to a construction project (Walker & Chau 1999).

Williamson (1981b) considers that the transaction cost approach has been applied at three levels of analysis. Firstly, at the level of the overall structure of the enterprise, which asks how the operating parts are related to one another: a direct reflection of the systems approach to organisation design. The second level focuses on the operating parts and asks which activities should be performed within the firm and which outside it and why. The third level is concerned with the manner in which human assets are organised to match internal governance structures.

This description complies with Coase's (1991) idea of co-ordination costs, which reflect the basic systems component of integration. The transaction costs involved in providing governance through the market will include the cost of drafting contracts, setting up the mechanisms for obtaining tenders, preparing other contract documentation and for ensuring the subsequent execution of the contracts by consultants and contractors. These are practical manifestations of the more abstract project management activities of integration, maintenance and monitoring.

The basic idea is that organisational variety arises primarily in the service of economising in transaction costs and that transaction costs are assigned to governance structures which differ in their organisational costs and competencies (Williamson 1985). This idea contributes significantly to an understanding of why a variety of organisational forms are used to manage construction projects by the same and by different clients. Whilst the transaction cost literature has been concerned with firms, the project organisation as a temporary multi-organisation (Cherns & Bryant 1984) also lends itself to similar analysis, as the economy of its co-ordination function should determine its form.

A client could define, design and construct a building entirely in-house (hierarchy) when both production and transaction costs would be internal to the client. Transaction costs would include the cost of identifying the project, setting up a design office and a contracting organisation, and hiring all necessary personnel and monitoring their performance. This is a task on which a client will embark rarely, that is only when transaction plus production costs are lower than providing these activities through the market.

Normally, clients will purchase the design and construction activities required from the market. The client will decide the type of building required and a strategy for providing it. The costs generated by this process are transaction costs. Advice may be obtained on contract strategy from a project manager (which may not be the best advice in terms of economising in transaction costs), which is again a transaction cost. The contract strategy selected will generate a governance structure. Williamson (1981b) recognises that the choice of an appropriate governance structure is pre-eminently an organisation theory issue.

As governance is the equivalent of the management system defined by project management theory then the process both within the client organisation and external to it comprises the governance structure and its transaction costs. The costs for activities within the client organisation, such as approving proposals, defining objectives, and providing policy directives, and the cost of hiring project management functions in the market, are all transaction costs.

The attraction of transaction cost analysis to project management is that it integrates economics, organisation theory, contract law and behavioural assumptions in an interdisciplinary study of organisational phenomena (Williamson 1981b). It adopts a comparative institutional approach in which the transaction is made the basic unit of analysis (Williamson 1985). Using the term governance structures to include the organisational approaches required to regulate and control activities, Williamson (1981b) generalises that in the long run governance structures that have better transaction cost economising properties displace those that have worse. This view lines up with systems theory in that organisations will survive only if they adapt to their environment. Members of the organisation make the decisions that adapt organisations to environments and which define governance structures, hence the behavioural characteristics they exhibit will influence the optimality of the structures chosen. Williamson believes the object is to identify the most economical governance structure and that it seems generally sensible that simple governance structures should be used in conjunction with simple contractual relations and complex structures for complex relations, reflecting contingency theory in relation to task and environmental complexity.

What are the transaction costs?

What then are the transaction costs which the client has to bear when developing a project? Essentially they are the costs of setting objectives, integrating contributions, making the various managerial decisions and controlling the contributors, costs that arise from the organisation structure selected, all of which are aimed solely at achieving the client's objective. The structure selected should aim to minimise transaction costs. Before this can be achieved, an understanding of how transaction costs arise is needed and is provided by project management theory.

The analysis here relies on the positivist version of contingency theory as adopted in earlier chapters and defended by Donaldson (1996). Whilst a number of reformulations have taken place, for example, strategic choice (Child 1972) and typologies (Mintzberg 1979), the deterministic mode still remains relevant to project organisation design as there is little or no scope for choice as the project organisation structure that fits the contingencies will need to be adopted to avoid underperformance of the project in use. It may well be that the task-specific duration-limited nature of the temporary management structures required to manage construction projects for clients is not susceptible to neo-contingency analysis. The organisation of the firms which contribute to project management may well respond to such analysis but promoters of such advances nevertheless state that 'the fact still remains that a basic contingency argument is inevitable' (Sorge 1991).

Analysis in these terms focuses upon the need to integrate the differentiated yet interdependent contributors. The construction process requires a high level of integrative activity that has not traditionally been recognised and provided. Also the control function should be designed to reflect the technical demands of the project and its environment and be based on the anticipated decision points that form the 'pinch points' through which the process must pass if it is to make progress. Decision points characterise the process and determine the interdependency of the contributors to each decision. The organisation structure for each project should be developed from first principles. Although a range of 'standard solutions' may emerge, it should not be presupposed that any predetermined solution is correct.

As discussed in Chapter 7, decision points can be classified into one of three types; primary, key and operational. Primary decisions are invariably taken within the client organisation and may not incorporate advice from construction consultants. The next level is key decision points. Key decisions are also made by clients but are usually based on the advice of construction consultants. A project team makes operational decisions upon which key decisions are based. The costs of arriving at operational decisions are production costs whereas the costs of the other decisions are transaction costs borne by the client. Contributors to operational decisions include the appropriate consultants and contractors. Project management has responsibility for integrating the project team to ensure that all advice has been given before an operational decision is taken. The tasks separated by operational decisions define the operating system, which comprises all the professional and technical tasks required to design and construct the project. The form taken by the units carrying out the tasks e.g. private practices, in-house to the client, can be explained by transaction cost economics.

The managing system carries out decision making, maintains, controls and regulates the operating system, controls the boundaries between the subsystems created by decision points and integrates their output to ensure that primary and key decisions are compatible with the clients' requirements. It also controls the boundaries between the process and its environment and between the process and the client and its environment. It monitors the performance of the subsystems to ensure that appropriate approaches and techniques are used, ensures that the resources producing the output of the subsystems (in particular, people) are procured and replenished. Also included are the activities of recommendation and approval of the propositions that arise from the system. Whilst many of these activities appear to be technical, they can only be achieved effectively if the behavioural characteristics of the contributors are channelled to the objectives of the project and to the benefit of the project outcome.

None of these activities is costless. They generate transaction costs, hence project management on behalf of a client is entirely a transaction cost. The managing system can be organised in various ways. It may be solely within a client's organisation: or it may consist of a client's representative and a 'consultant' project manager, or the architect acting in a dual role of manager and designer. Transaction cost economics argue that economising in such costs takes place to determine how such activities are organised. Whilst some costs may be easy to conceive, others are not; the costs of approval and recommendation can be hidden, but large, as they frequently involve major groups of high-level people in client organisations.

The above articulates the transaction costs of project management using project management terminology. In contrast, Williamson (1985) as a transaction cost economist, describes transaction costs as drafting and negotiating agreements, set-up and running costs of the governance structure which monitors and settles disputes, haggling costs, and bonding costs of effecting secure commitments. Dahlman (1979) classifies transaction costs as search and information costs, bargaining and decision costs, policing and enforcing costs. He then argues that the three classes reduce to one: resource losses due to lack of information. Reve and Levitt (1984) presented one of the few attempts to relate these ideas in practical terms to construction, but their focus was on the client/contractor relationship not project management. Nevertheless the compatibility of the ideas is clear.

The recognition that economists had largely ignored management was the catalyst for integrating economics and organisation theory. Coase (1991) said that 'what happens in between the purchase of the factors of production and the sale of the goods that are produced by these factors is largely ignored by economists' and that 'what economists have conventionally studied is a system which lives in the minds of economists but not on earth'. Demsetz (1991) acknowledged that understanding of firms could be improved by recognising that management is a scarce resource in a world in which knowledge is incomplete and costly to obtain. This will not startle students of management but the fact that economists seem to have only recently discovered this helps to explain the lack of relevance of much economic theory felt by managers.

Many transaction costs arise from preparing and enforcing contracts and procedures designed to produce behaviour on the part of members of project organisations that is compatible with project objectives. However such processes are not foolproof, hence there is a risk that such behaviour is not compatible and not controlled, the cost of which can also be conceived as a transaction cost. Partnering is a concept which aims to overcome such problems as discussed in Chapter 6.

Production costs

Many accounts of transaction costs fail to make clear that the object is to economise in the sum of transaction and production costs. As Demsetz (1991) puts it; 'the emphasis that has been given to transaction costs (or that has been claimed to be given) dims our view of the full picture by implicitly assuming that all firms can produce goods or services equally well'. Economists' generalisations of transaction costs equate to project management activities. Specialist consultants, e.g. architects and engineers, are producers and their costs are mainly production costs although some of their functions may be categorised as transaction costs, e.g. architects who also manage the project. Quantity surveying is a transaction cost. An increase in transaction costs may be accompanied by a reduction in production costs but this does not have to be so. Each case has to be judged on its merits.

These ideas throw up some interesting questions from the client's point of view. The client, whether private or public sector, wishes to acquire a project in order to satisfy a need which is to do with the client's main purpose which may be, for example, to deliver health care or to manufacture motor cars. The client has a wide range of organisational structures available for designing and constructing the project. These structures can provide the various skills and commercial activities either through the market, in-house or in some combination of these. Two basic questions arise. How are costs (of either kind) defined in construction? Which are transaction costs and which production costs?

The view that emerges from the transaction cost literature is the acquisition in the market or hierarchy of a product that can be clearly defined and delivered in the form specified, and that either the market or the firm can deliver it in exactly the same form (albeit perhaps at different production costs). This is not the case in construction where the product is usually bespoke and would be designed differently by all designers and as a result would perform its function to different degrees of effectiveness. The extent to which the project meets the client's expectation will be a product of the definition of the client's requirements, the effectiveness of their articulation by the designers and the effectiveness of the realisation of the design. The methods by which these processes can be monitored and controlled are numerous and generate different costs. Also, production costs can vary relative to the quality of the product. For instance a world famous architectural practice may well charge more than a local one. The quality of the output will vary (but not predictably). The idea of cost in relation to construction projects then becomes most complex.

The second question asks which are transaction costs and which are production costs. This question revolves around the costs of project management functions. Even when project management and all other skills and activities are obtained from the market, some management costs will arise in the client organisation as a result of the firm embarking on a construction project. The converse will also apply in that even when the client purports to undertake the project entirely in-house some elements are likely to have to be acquired from the market. This reflects Reve and Levitt's (1984) argument that market and hierarchy overlap. Their identification of trilateral governance and Stinchcombe and Heimer's (1985) conclusion that contracts for construction incorporate aspects of hierarchy in order to protect against opportunism, are supported. In this respect, Cheung's (1983) generalisation of Coase's theory of the firm is relevant. Coase's main concern is the choice between (complete) direction by the market and (complete) direction by the firm. Instead of viewing the market and the firm as competing rivals, Cheung suggests that what matters is the choice of contractual arrangements.

Behavioural assumptions

Williamson (1981b) believes that two behavioural assumptions underpin the transaction cost approach; bounded rationality and opportunism, and that these assumptions represent 'human nature as we know it and supplant[s] the fiction of economic man'. It is worth examining these ideas relative to the project management process.

Bounded rationality implies a limit on rationality, not in terms of being 'partly irrational', but in contrast to the traditional assumptions of economics of the perfectly rational being, in that people act rationally but are limited by their analytical and data-processing capabilities. This is a concept that is easier for managers to accept. It is reflected in clients' project management by the likelihood of their not selecting the most appropriate organisational structure and procurement method, usually as a result of not considering the full range of options available to them due to inappropriate early advice. This is less likely to be the case for clients that build regularly.

The second behavioural assumption of opportunism is more controversial. Opportunism takes a dismal view of human nature as it maintains that 'human agents will not self-enforce promises but will defect from the letter and the spirit of an agreement when it suits their purpose' (Williamson 1985). Accompanying terms are moral hazard and shirking. As referred to earlier, the conjunction of economics and organisation theory has led to much debate due to the different human motivation and behavioural perspectives taken by the two disciplines. Managers object to the assumption that they behave opportunistically, whilst economists argue that the cost of distinguishing between opportunistic and non-opportunistic behaviour is in fact a transaction cost. Williamson's (1985) view of opportunism is that 'this unattractive view of human nature nevertheless generates numerous refutable implications' and 'does not preclude the possibility that they [individuals] will forge durable alliances'. He also argues (Williamson 1990) that 'organisation theorists were familiar with opportunism long before economists got around to it'. Thus in undertaking a transaction cost analysis, behavioural theory is incorporated into the analytical framework, albeit from an economic perspective, and other models of organisation can be used to explain departures from the assumptions.

Both Winch (1989) and Reve and Levitt (1984) recognise the potential for opportunism in construction. Winch uses the illustration of change in project specification leading to opportunistic pricing of 'extras' by contractors. Reve and Levitt argue that clear relationships between contractors and consultants reduce incentives for opportunism. Detailed documentation of the contracts between clients and main contractors and between main contractors and subcontractors, e.g. specifications, are intended to prevent opportunism. The cost of their preparation by consultants is a transaction cost. Whilst these costs may be high, the potential for high hidden costs (or losses) due to underperformance of the completed project due to opportunistic behaviour by consultants is even greater. Contracts between clients and professional advisors, including project managers, are not usually as carefully prescribed as between clients and contractors, providing scope for opportunistic behaviour by consultants. This is further complicated by the fact that professional advisors' efforts are difficult to measure and information is distributed asymmetrically between the client and the professional advisors.

Williamson (1979) recognised that transaction costs are particularly significant when economic agents make relationship-specific investments (asset specificity) arising in three ways: site specific; plant specific; and human asset specific. In situations such as these, buyer and seller are locked in, competition is before investments are made but not afterwards. Asset specificity raises opportunities for opportunism.

The focus of asset specificity in construction due to site and plant has been on the relationship between the client and contractor but not on the project management process. This has led to the ideas of bilateral (Williamson 1979) and trilateral governance (Reve & Levitt 1984) as a function of frequency of investment and specificity. Winch (1989) believes asset specificity due to site and plant will probably only relate to particular types of civil engineering projects. Human asset specificity is more widely relevant to project management where, as the design develops, detailed knowledge is held in a firm, usually by a relatively small number of people. Both client and consultant are locked into the arrangement (Reve & Levitt 1984).

In such circumstances, the impact of transaction costs relative to project managers, designers and other consultants is interesting as detailed contractual arrangements may not be made, rather the relationship relies on trust and the professional standards of the consultants. In such circumstances the prospect for opportunism is high as many clients do not monitor the performance of their consultants. Whilst the level of actual transaction costs may not be great, if the completed project performs at less than the expected effectiveness the cost of not monitoring effectively can be extremely high. Alternatively, and increasingly, detailed negotiations are carried out with consultants before appointment, contract conditions are spelled out in detail and monitoring is undertaken, leading to high transaction costs that in the long term may be advantageous. Increasingly therefore reliance on formal contractual relations emerges, not only between clients and contractors but also between clients and project managers and consultants which lends the process amenable to analysis as a nexus of contractual arrangements (Cheung 1983, 1992). With increasing commitment to project management by clients, the recognition of human asset specificity related to design knowledge may move the relationship to a bilateral monopoly. A similar relationship may also exist between the client and contractor when relational contracting, e.g. partnering, is found to be valuable in reducing transaction costs.

In practical terms

Project management is not costless. The project management costs carried by a client are entirely transaction costs and the role of project management is to minimise the total of transaction and production costs. For clients to make the best decision regarding the project process to be adopted, they need to consider their internal costs in setting up the project, the cost of any external project management and of the consultants and other advisors and the production costs of designing and constructing the project. Frequently, in practice, only construction costs, which equate with the predicted tender sum, are considered. Reducing transaction costs at the expense of a greater increase in production costs is obviously unsound but is not a factor that is explicitly considered for most projects. Against the total of transaction and production costs, clients have to consider the extent to which the completed project fulfils their need; aesthetically, functionally, financially and in terms of delivery. Minimisation of total costs using a system that does not deliver a project of an appropriate quality does not maximise the client's benefit. This process will always carry some element of risk as, when selecting the process for producing the project, the output cannot be entirely guaranteed. On a micro level this can be illustrated by the unlikelihood of it being possible to develop specifications and contract documents to such an extent that additional unanticipated costs are not incurred. Incurring higher transaction costs to reduce the likelihood will not eliminate the possibility, so the issue becomes one of the amount of risk to be accepted by the client against increasing transaction costs.

Put in practical terms, the significance to construction professionals of an understanding of the relationship between transaction cost economics and project management theory is that:

- When recommending project organisation structures to clients the total cost of project management (transaction costs) and design and construction (production costs) should be evaluated.
- Different organisation structures generate different project management approaches and transaction costs which in turn generate different costs for design and construction.
- Higher project management costs may not lead to lower design and construction costs and vice versa.
- The choice of organisational structure should minimise the sum of project management and design and construction costs whilst delivering a project that meets the client's requirements.
- Organisation structures which increase project management costs can only be justified if design and construction costs are reduced by a greater amount and the effectiveness of the project is maintained or, if design and construction costs are held constant, by improving the effectiveness of the project.

The project management costs to be minimised in such an evaluation include:

- In-house client costs such as the opportunity costs of the time that senior executives spend on the project; a major part of which comprises defining objectives, their detailed development and the cost of making decisions.
- The cost of establishing the project organisation structure, which may also require substantial input from the client's executives.
- The cost of integration within the client organisation and within the project team (including the contractors) and between the client and the project team.
- The cost of producing contract documentation for agreements with consultant and contractors, including terms of engagement, contract agreements, bills of quantities and specifications.

- The costs of negotiation between all parties; client, consultants and contractors.
- The cost of monitoring agreements and contracts with consultants and contractors to ensure that all parties are complying, including supervision of construction work.
- The cost of enforcing contracts, including all dispute and settlement costs.

Project management theory explains the choice of an organisational structure as an interaction of the environment and the project tasks and identifies the activities that constitute project management on behalf of a client. While such generalisation is useful in terms of providing a general framework for analysing how decisions are made and projects managed, it is sometimes too general for analysing real world problems. While some detailed description of the major features of the process is necessary to operationalise the analysis, the problem is that of identifying fundamental issues. Moreover, there is also the danger of giving too detailed a description which results in ad hoc theorising leading to results with no general applicability. The transaction cost economics approach can be used to supplement project management theory by focusing on the market forces that determine the behaviour of the parties involved in the system. Thus the transaction cost economics approach not only provides an explanation for the choice of project management system, it also provides insights to the fundamental elements that shape decisions in the development process. Most importantly it can also give more rigour to organisation theory by providing the analytical framework for identifying hypotheses for empirical work. On the other hand, the application of the seemingly reductionist transaction cost economics approach to real life situations can be improved by richer specification of the management process which determines the actual transaction costs incurred and their distribution amongst client, consultants and contractors. For practitioners, the theoretical ideas provide a framework for deciding which project management system to recommend to clients. For clients it offers a framework for critically reviewing the advice offered to achieve a more objective evaluation than is usual. It lays down a basis for learning so that mistakes are not repeated.

The components of project organisation structures

In practice there are three major components to the organisation structure of projects:

(a) The client/project team integrative mechanism.
(b) The organisation of the design team.
(c) The integration of the construction team into the process.

A number of options are available within each of these categories and this results in a large number of possible combinations. The whole range cannot be considered here but the more likely alternatives are analysed. Of course some options within (b) would not be used with some options within (c), e.g. a traditionally organised design team with a design-and-build contract. However, many others are combinable, and have an effect one upon the other, e.g. a traditionally organised design team and management contracting, which could have a profound effect upon how the design team is organised.

Client/project team integration

This aspect has been discussed earlier, and, as was pointed out, the variety of organisation structures of client organisations is vast. It will not be possible for the project team to affect the client's organisation structure other than marginally. It will be up to them to organise themselves to fit in with the client's organisation. It is therefore a case of the project team organising itself so that it has the capability to understand the client's firm and its environment in order that it can respond to the client's requirements and any changes that may be dictated by the client's environment during design and construction.

It will be easier to integrate with some clients than with others. Where the client has in-house expertise in construction, it is to be expected that the dovetailing of the project team with the client will be easier. On the other hand, integration with a client who does not have in-house expertise or, even more difficult, one who has not built previously, will be more of a problem. The response therefore has to come from the project team and the structure of the design and construction teams should be set up so as to reflect the difficulty of integration with the client. For example, if the client is experienced in construction and has in-house expertise, the client may appoint a project manager with experience of the construction industry from the client's own staff. In such a situation, given the right qualities in the project manager, it may well be that a traditionally structured design team under the direction of the project manager would be appropriate and economical. There could well be no case for the appointment of a further project manager from the design team.

Alternatively, the client may be building for the first time and embarking upon a complex project, e.g. the rebuilding of a processing plant. Because of the naivety of the client and the complexity of the project, it may be advisable to appoint a project manager in an executive capacity from outside the client's organisation.

Similarly, as referred to previously, the authority of the project manager will vary depending upon the attitude of the client to delegation. The extent of delegation of authority is likely to be strongly influenced by whether the project manager is 'in-house' to the client's organisation or external to it. There will be a tendency for clients to delegate more to an 'in-house' project manager and such a project manager will also have greater access to the internal workings of the client organisation. Even in such a case delegation may not be high if the client's organisation is hierarchical and bureaucratic.

There can be no hard and fast rules for the integration of the client and the project team. The mechanism that is selected should be the result of an analysis of the client's organisational structure, the client's needs and the complexity of the project. The objective of such a mechanism should be simplicity within the constraints of the need to identify clearly decision points and the client's involvement with the decision-making process.

Design team organisation

Traditional structure

The traditional structure of the architect being responsible for design and management with other consultants acting for the architect and with estate management functions being directly responsible to the client is illustrated diagrammatically in Fig. 10.1. In such an arrangement the contractor is normally appointed after the design is substantially complete, usually by competition, although the contractor may be appointed on the basis of a negotiated tender or by some other means.

In many cases each contributor will be from an independent professional practice, the contractor also being independent of the other contributors, yet the contributors will be interdependent in terms of the project. The more complex the client organisation and/or the project, the more interdependent will be the tasks to be carried out in achieving the project and the more the contributors will rely upon each other to carry out their work satisfactorily.

Such a structure produces a high level of differentiation between the contributors, which demands a high level of integration. The problem of providing the appropriate level of integration is compounded by the fact that the managing system is not differentiated from the operating system. That is, the architect is attempting to fulfil dual roles. One is in the operating system – design – the other

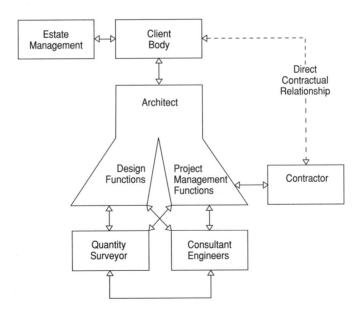

Fig. 10.1 Traditional structure.

in the management of the project. There is therefore a high potential for someone in this position not to be able to exercise objectivity in decision making. In addition, whoever is in this position is placed under severe pressure by being required to undertake tasks that frequently require what are often incompatible skills – design and management. This does not mean that adequate project management cannot in any circumstances take place in such a structure, only that it may be extremely difficult to achieve and will require a person of exceptional talent to be able to fulfil both roles satisfactorily in the complex environment within which construction takes place.

Although this does not have to be the case, such a structure has a tendency to restrict access of the other contributors to the client and hence the decision-making process. The perceived personal relationship between the client and the architect, particularly with clients new to construction, can inhibit the client from approaching the other contributors for direct advice. As there is no one solely in a project management role, there is a danger that apposite advice is not taken, which will be to the detriment of the outcome of the project.

Integration within the design team can therefore be difficult to achieve in this structure, as can the integration of the design team with the client. This situation is made even more difficult by appointing the contractor in competition. The direct contract with the client which this produces reinforces the contractor's differentiation from the design team. Similarly, the frequent exclusion from the design team of the estate manager also adds to differentiation. Whilst the traditional system can be seen to be problematic, it has been argued that this conventional approach may still be just as effective as more novel approaches (Bresnen 1991) and that such approaches are often re-selected in preference to the uncertainty and disturbance that may ensue with a departure from normal practice (Bresnen & Haslam 1991). The latter is, of course, only the case when the client is experienced with the system.

The situation can arise in which the quantity surveyor, estate manager or engineer is the first contributor to be appointed and subsequently has advised the client on the appointment of the other consultants. In each of these cases the result will tend to be similar to the situations described in Fig. 10.1. That is, the first appointed contributor assumes project management responsibilities alongside the professional functions, leading to a potential lack of objectivity in weighing factors from other contributors and to integration problems equally as difficult as those described above.

The degree of differentiation would be reduced if all the design team contributors were from the same interdisciplinary practice but, even within such a practice, if its members were organised on conventional lines with the team leader exercising both professional and project management functions, the main hindrance to objectivity and integration would still remain. However, such a practice has a better opportunity of overcoming problems created by differentiation and of generating sound integration for individual projects than if projects are designed using independent practices. A parallel situation would exist if all the design contributors were in-house to the client's organisation (e.g. a government department, a local authority or a large industrial concern). A major additional advantage in this situation would be the potentially high level of integration with the client, as client and design team would be under the same organisational umbrella.

Taking this argument a stage further, the organisation that should, theoretically, have the least differentiation and the greatest opportunity to achieve full client integration is one that has a construction capacity as well as a design capability within the client organisation, for example a local authority direct labour organisation or a developer/contractor. However, problems of control and motivation have been shown to exist in the case of direct labour departments whereas developer/contractor organisations have been successful in many cases.

Non-executive project management

A structure often employed by interdisciplinary practices, either private practices or in-house to the client's organisation, is one that includes a non-executive project manager (Walker 1976), sometimes called a co-ordinator, who operates in parallel with the other contributors, as illustrated in Fig. 10.2. The role undertaken by the person in this position is based upon communication and co-ordination activities and is not concerned with decision making. In these circumstances responsibility for the success or failure of the project will be with the firm or the particular in-house department and not with the non-executive

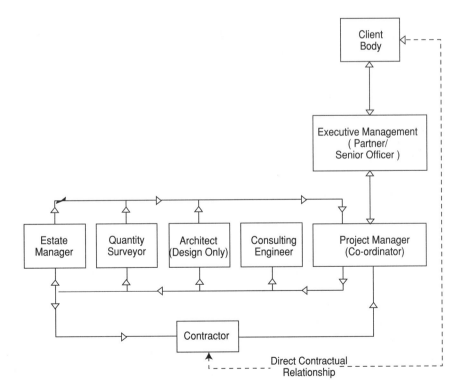

Fig. 10.2 Non-executive project management structure.

project manager within the team as would be the case with a project manager acting in an executive capacity. There is therefore less pressure for the project manager or the firm to define the project manager's role and authority. What pressures there are will be internal to the firm or departments, depending upon how they see the role of the project manager within the team.

Such a role is unlikely to have a significant effect upon the quality of integration of the design team with the client's organisation but could, if exercised with skill and received positively by members of the design team, assist in integrating the design team. If exercised unskilfully or in an unco-operative climate, it could emphasise differentiation within the team without contributing to integration. The authority of the non-executive project manager is likely to be weak and hence his or her ability to contribute will be determined by the commitment of the firm, his or her informal authority, and the attitude of the individual members of the design team to his or her role.

A non-executive project manager could be used where the contributors are from separate practices. Such a role is likely to be undertaken by one of the independent firms contributing to the design, although it is conceivable that it could be undertaken by a firm solely devoted to project management. The result is unlikely to contribute much to the project management process, although any improvement in co-ordination and communication would be of benefit. The lack of an executive function would mean that project management would be split between the non-executive role and the decision-making role, which would still be undertaken by the architect, partner of the lead consultancy or senior officer in the case of a public authority. The potential benefit of improved co-ordination and communication may well be more than offset by the complexity of the management system which emerged. The major management role will still not be separated from the operating system and, in fact, if the non-executive role were shared with an operating role, this situation would be further compounded.

Executive project management

An executive project management role (Walker 1976) is undertaken by a firm or person independent of the other contributors to the process, as illustrated in Fig. 10.3. Similarly, if the design team is part of an interdisciplinary practice or in-house to the client, it will be undertaken by a member who is also independent of the other contributors. In such a structure the project management activity occupies a dominant role in relation to the other contributors, and although they operate as a team, the project manager will make the decisions that are within the purview of the contributors. He or she will tend to be the sole *formal* point of reference to the client for the purpose of agreeing and transmitting the decisions that must be made by the client. In addition, the project manager will be concerned with controlling, monitoring and maintaining the project team, as discussed previously. These activities are far more dynamic and purposeful than the co-ordination and communication activities of the non-executive project manager and do, of course, subsume them.

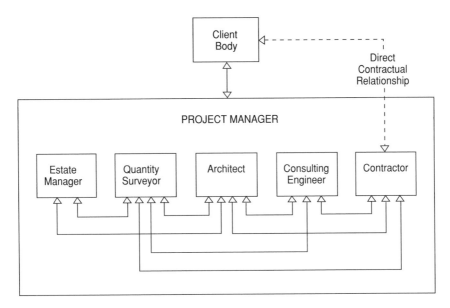

Fig. 10.3   Executive project management structure.

It is necessary that the firm or person undertaking this role ensures that re-sponsibility and authority are clearly established with both the client and the other contributors to the project. Although this is difficult to achieve, the ben-efits of producing a situation in which the roles of the contributors are clearly established are significant. It is, of course, unrealistic for the project manager to accept responsibility for the technical work of the specialist contributors, but responsibility for progress and for budget control are possible given the author-ity to report back to the client if contributors are not performing satisfactorily. However, in practice, the project manager should work in a collaborative man-ner with the contributors and a major role would be one of facilitating the work of all the contributors so that the project is developed by a team approach. The project manager's primary concern would be that appropriate decisions are taken by both the client and the project team at the right time.

This is not to say that the use of a project manager independent of the other contributors is without problems. In a survey of project management companies in New Zealand (Wilkinson 2001), it was found that 'lack of contact with the client could be a source of grievance for the other professionals, which often transferred into frustration with the project management companies'. The role and responsibilities of the project management company were seen as needing to be made clear to all members of the team. Whilst communication was seen to be improved, there were complaints about the project management companies' inability to listen to other professionals in the team.

In an interdisciplinary practice or in-house to the client, the responsibility for the project is clearly with the firm or the department, respectively. The authority of the project manager will be decided internally in the case of an interdisciplinary practice and it is unlikely that the client would exert as great an influence as it would over an in-house department.

The major benefit claimed for the executive project management structure is that management becomes clearly separated from the operating system. That is, no person is charged with carrying out both design activities, be they architecture, engineering or quantity surveying, and project management activities. This allows concentration upon the management needs of the project and makes it possible for conflicting professional advice to be considered more objectively so that decisions which are in the best interests of the project as a whole can be made or recommended.

The structure facilitates integration with the client because the person responsible for managing the project within the client's organisation can readily identify the management responsibility within the design team and is likely to have empathy with the person in this position. This should facilitate the decision-making process, particularly within the client's organisation. A major role of the project manager would be the planning and programming of the project, which would include identifying the contributors and their roles. In doing this the project manager should recognise the differentiation generated by the particular project. The project manager should be in a position to judge the integrative demands and should design mechanisms to cope with them. It is to be expected, therefore, that the project manager will be intimately involved with the client in determining how the organisation for the project should be structured and in deciding the firms and people who will undertake the various roles.

In certain circumstances the client's organisation has personnel who can undertake the executive project manager's role without the need to appoint someone from outside the client's organisation. This arrangement can contribute significantly to the ease of integration between the client and the design team, provided, of course, that the project manager has sufficient status within the client's organisation and can command the respect of the design team. If not, there may be a tendency for the design team to bypass the project manager and seek higher authority in the client's organisation. In this case the role of the project manager would be seriously undermined with resultant confusion in the decision-making process. A similar situation can arise when the project manager is outside the client's organisation, and the solution will be in the attitude which the client takes to resolving the situation.

The person appointed from within the client's organisation to liaise with the design team when an executive project manager has been appointed from an outside organisation is also often called a project manager and it is therefore important to recognise their different roles.

It is also important to recognise that if a person in the design team who is a member of one of the firms contributing in a professional capacity (e.g. quantity surveyor, engineer) is given the title of project manager and ascribed specific responsibilities and authority in this capacity, this does not constitute an executive project management structure as described here. It is in fact a variation on the traditional structure as they will be acting in the same way as the architect traditionally acts, and some of the objectivity of the executive structure would be lost.

In the early 1980s the British Property Federation (1983) devised a system, which was not dissimilar from the executive project management approach, in response to their perceived need to improve the traditional structure as it had become increasingly concerned about 'problems of poor design, inadequate supervision, insufficient choice of material and contractual methods which caused delay and increased cost'.

It restructured the traditional organisation and hence relationships between the participants to projects by introducing a client representative and a design leader. The client's representative was akin to a project manager and 'assists the client to develop the concept of the project; he advises on the selection of consultants, he resolves conflicting priorities, instructs the consultants and contractors and safeguards the client's financial interests'. The client's representative may be from within the client organisation or an appointed individual or professional firm from outside the client organisation.

Design consultants were responsible for all aspects of design, e.g. function, structures, services and cost, but they were co-ordinated by a design leader who had overall responsibility for all aspects of pre-tender design and specification. Although the system laid down the individual responsibilities of the client representative and the design leader, the duties were transferable between these managers. The client representative was responsible for all aspects of management during construction but worked in conjunction with the design leader and a supervisor. The supervisor, who was appointed by the client, monitored the work to ensure the building was constructed as designed and specified in the contract documents.

It was claimed that the BPF system was innovative and challenging to the professions and industry. In terms of its organisational structure its innovation lay in the introduction of additional levels of management. This development subscribed to the idea of an open system moving towards a higher level of organisation with greater differentiation and a corresponding need for greater integration. The system also had the attribute of flexibility so that it could be tailored to suit the particular project to which it was applied.

In fact, the most dramatic innovations of the system were in its procedures rather than in its organisation structure. Many of the features of the organisation structure had been recognised for some time. The novelty was in the manner in which they were amalgamated into the system and integrated with the more innovative procedural initiatives such as an adjudicator to settle disputes, fixed fees plus incentive fees for consultants, elimination of the bill of quantities, contractor designs, the use of prices linked to construction activities and the elimination of nominated subcontractors. Whilst not in wide use, the system's ideas have had a strong impact and influenced the development of project management generally.

Integration of the construction team

The degree to which the construction team can be integrated into the process at the design stage is determined by the tendering arrangements which are made for obtaining the price for construction. This is obviously a key decision as it has a fundamental effect upon the organisation of the whole process. The most common procurement method is by a lump sum competitive tender after the design has been substantially completed. This traditional system continues to dominate, not only in the UK but also elsewhere (Love *et al.* 1998; Bowen *et al.* 1997). The principal reason for this appears to be familiarity with the system rather than an objective selection of the most appropriate method. A number of other methods are available, which provide for more constructive integration of the contractor and are increasingly being used, particularly in the UK.

Competitive lump sum tendering after completion of the design provides the least opportunity for integration as this method requires that the contractor who is to construct the project cannot be involved in the design stage. In addition, during construction it is often difficult to integrate the design team with the construction team as the split between design and construction appears, in many cases, to create a psychological barrier between the two groups. The contractor will often feel that the design has been carried out by people who do not understand construction methods and who seem to be providing the wrong drawings at the wrong time. The designers may adopt the view that the contractor is only concerned with profit and not with providing the service that will provide the project they require. Whether such views are correctly held or not does not overcome the fact that on conventionally organised projects, the greatest degree of differentiation occurs between the designers, normally represented by the architect, and the contractor.

If the project is managed by an executive project manager, then the greatest integrative effort is likely to be centred around this interface. If the project is organised conventionally, problems of integration at this point will be extremely difficult to resolve if the architect is both designer and project manager and has to also integrate with the contractor. The difficulties of integrating this interface are compounded still further by the use of subcontractors, both nominated, named and domestic. Subcontractors nominated or named by the architect will have a strong allegiance to the architect while having to work in a contractual arrangement and under the direction of the contractor. Domestic subcontractors hired directly by the contractor will tend to hold the same views as the contractor towards the designers.

Many of the recently developed methods of appointing contractors have been aimed at allowing them to be better integrated into the design team while still allowing an element of competition in obtaining a price for the project.

Two-stage tendering

In order to maintain competition in a similar form to the traditional method, yet allow the contractor to be involved to some degree in the design stage, two-stage tendering emerged in the late 1960s but has not developed to any great extent. In the first stage, selected contractors are invited to tender. Their tenders are based on a notional bill of quantities in which the items are fully described and the quantities are hypothetical but of the order of magnitude anticipated in the proposed project. The successful tenderer is then involved in the further development of the design as a member of the team.

A bill of quantities is prepared for the fully developed scheme and is priced by the successful tenderer using the rates, where applicable, in the first-stage tender and negotiating other rates on the basis of the original tender. The result of this process is the price for the project.

This approach assists in the integration of the contractor into at least part of the design process but does not fully exploit the potential benefit to the project of involving the contractor from the beginning. Perhaps one of the greatest benefits is the opportunity to involve the contractor in selection of the specialist subcontractors.

Designers are often sceptical about the contribution a contractor may make to the design of a project but this is one of the aspects the project manager must overcome in integrating the contractor into the design stage. The advice the contractor can give regarding the constructional implications of design decisions and construction methods and processes is likely to be recognised by the designers only after the event. This makes the project manager's task that much more difficult. It also illustrates the fact that integration of the contractor into the design process is less likely to happen in the traditional process where the decision to integrate would have to be made by designers.

Two-stage tendering represents a 'trade-off' between integration of the contractor into the design stage against a traditional approach to competition. It was one of the earlier approaches to integration of the contractor. However, it has been found 'that the "two-stage tender" procedure is insufficient to ensure team integration, collaborative work and efficient achievement of goals in construction projects' (Cicmil & Marshall 2005).

Serial tendering

Serial tendering is used to obtain tenders for a number of similar projects. Contractors bid on the basis of a notional bill of quantities and normally the lowest is accepted. The prices in the notional bill are used for a series of projects, the number, timing and size of which are indicated to the tenderers before bidding.

The actual price for each separate contract is calculated by using the rates submitted in the notional bill. On the face of it, this procedure allows the contractor to be integrated from the beginning of the design of each building in the series for which the contractor has been successful. However, this is rather artificial as the majority of the design decisions will have been made before production of the notional bill so that although the contractor may be integrated, the contractor's effective influence on the design stage is limited. Nevertheless, there is a distinct advantage over the traditional approach as it allows discussion with the contractor about such things as subcontractors, plant, programming, etc. during the design of each project in the series.

Negotiated tenders

The use of negotiated tenders does not rely upon a competitive element in selecting the contractor. The contractor is selected on the basis of reputation and will probably have worked satisfactorily for the client and/or design team previously. The price for construction will be agreed with the contractor following negotiation between the quantity surveyor and the contractor. There are a number of variations in the approach adopted, often including some types of target cost. Whichever approach is used, the effect on the organisation structure for the project will be similar. Under this arrangement the contractor can be involved in the design process at whatever point the client or project manager decides.

Integration of the contractor can achieve the highest possible level using this approach, but on many negotiated contracts the contractor is still not brought into the process until the design is well advanced and one of the major advantages of this approach is lost.

Naturally, such an arrangement requires a high level of trust between the client, design team and construction team. It is often said that a negotiated price will be higher than a competitive one so the integration of the contractor may be traded off against a higher price. However, the benefits to be gained from integration – earlier start on site and earlier completion, constructionally sound designs, cost-saving advice from the contractor, etc. – may more than counterbalance the lack of competition. Within this framework, subcontracts may be negotiated or competitive, giving a facility for closer integration of subcontractors if their use is considered beneficial.

Management contracting

This approach is fundamentally different from those previously described, as it is a method of integrating contracting expertise into the design stage of a project on a fee basis rather than a method of obtaining tenders and letting contracts. The objective is to incorporate the management contractor into the design team on an equivalent basis to all the other consultants. The management contractor would be responsible for the construction work, all of which is carried out by subcontractors who may be appointed in competition or by negotiation.

The project is split into 'packages' for tendering purposes. Design, cost control and construction are integrated into the overall programme. The contractor provides advice on the availability and procurement of materials and components and the 'buildability' of the proposed design. The management contractor is involved in the compilation of the cost plan for the project and monitors and provides financial data concerning the project during the construction phase. The management contractor does not carry out the construction work directly but may provide certain central facilities (e.g. canteen, welfare, scaffolding). However, the management contractor sees its role as that of a manager.

It can be seen that management contracting is a positive approach to the integration of construction expertise into the design process. Its main trust is management of the construction aspects of the project in both the design and construction phases with status equivalent to that of all other professional contributors.

Whereas the subcontractors will be in a strongly differentiated position similar to that of the contractor in the traditional process, the management contracting activity will be primarily concerned with integrating them. This is rather different from the situation with a traditional structure as the management contractor's objective is that of achieving satisfaction for the client and he or she does not have the entrepreneurial interest of the contractor employed under a conventional arrangement. This is in fact transferred to the subcontractors but in this case there is an integrating mechanism acting directly for the client.

The person responsible for project management will have the task of integrating the management contracting activity into the design team. If this is the architect, then it may be difficult, although it should not be so if the architect instigated the use of management contracting in the first instance. However, frequently it is the client who decides that management contracting should be used. In this case it is important that the remainder of the team is structured in a compatible manner. The allegiances and attitudes at large in the construction industry may make extremely difficult the integration of management contracting into a design team whose other members are unsympathetic to the idea. This may mean that either a strong client or a project manager with sound authority is necessary to gain greatest benefit from this approach.

Separate trades contracting and construction management

Separate trades contracting is used as a generic title for approaches that can be seen to be similar to the idea of management contracting. In each case there is no main contractor appointed for the project but instead a member of the client's organisation, a project manager external to the client's organisation or fee-earning members of the project team (e.g. the architect) organise subcontractors to undertake the work. The idea is not new, although it is often claimed to be in its new guise of construction management (Hughes 1997; Chartered Institute of Building 2002). Scotland was the last region of the UK that used it as the normal method of arranging the construction stage. In the construction management version, which is currently increasing in use, the client contracts directly with a series of trade contractors and assumes many of the functions of a general contractor. The system ideally requires that the client has a competent in-house capability in project management as not only has the client to organise the separate trades contractors but also has to integrate with the designers and other consultants. Hughes points out that its main proponents argue that it is more a new management philosophy than a system of contracting which is organic rather than hierarchical. In this respect it reflects the relational characteristics inherent in partnering. An interesting, if historic, form in the UK was documented by Thompson (1978), referred to as *alternative methods of management* (AMM), in which the site architect/manager was responsible for running the site and directed the activities of subcontractors either through their supervisor or directly to the men on site. Although there was no main contractor, there may be a need for a general builder to work alongside specialist contractors and to provide some central services. The subcontractors were appointed by competition or negotiation for packages of work. In effect the site architect/manager replaced the main contractor's site agent and provided the site with direct and constant design supervision. The site architect was often supported by a quantity surveyor who arranged the contracts and must have ready access to the client whose involvement on site was higher than on traditional projects. It was claimed that the main advantages were that communication was as direct as possible from client to architect to tradesman, that the human element was all important and the client's interest was best served by people committed primarily to the client's project rather than their profession or trade. Compared with management contracting, AMM does not incorporate construction expertise as such in the design stage but relies upon the ability of the architect in this respect.

Other separate trades contracting approaches are 'cost plus contracts' of various types. They are basically similar to the above, except that subcontractors are reimbursed their actual expenditure plus profit, often only within set total cost targets. The management implications are similar to those outlined above and demand similar levels of involvement of designers in managing the construction stage. They have not received the detailed attention in recent years that management contracting has, as they are really contractual arrangements and not approaches to management. They are rarely used except for emergency works as they can be expensive and can, of course, also be used for general contracting as well as for separate trades contracting.

Design-and-build

Design-and-build contracts are arrangements that do not separate design and construction as one firm offers the total package of design and construction.

Akintoye (1994) identifies six variations of design-and-build. Traditional design-and-build is defined as arrangements in which the contractor accepts total responsibility for both the design and construction of the project. Other types have varying degrees of involvement of the contractor with management of the design process and involvement in the actual construction of the project, including package deals and turnkey contracts which can be even more all-encompassing than traditional design-and-build. Worthy of particular note is 'novation design-and-build' in which the client's consultants, who have developed the project to the point of appointment of the design-and-build contractor, are passed to the contractor for the completion of the project. The opportunity to provide effective integration of the process is theoretically higher in design-and-build approaches than in more traditional methods although using the novation method presents less opportunity than traditional design-and-build.

In spite of the increasing client-led use of design-and-build, architects continue to be reluctant to embrace the system (Akintoye 1995). Of the architects surveyed, 20% of their private sector workload and 5% of their public sector workload was derived from design-and-build. Even so, the use of design-and-build appeared to give rise to strong sentient responses, which contributed further to differentiation within the system, not only between architects and contractors but also between architects and clients.

There is a longer experience with design-and-build for civil engineering projects and many contractors which specialise in civil engineering work have strong in-house civil engineering design capability. There is also a long history of civil engineering projects giving the contractor an opportunity to submit an alternative design for all or part of the project to allow contractors to increase the competitiveness of their bids. The discussion which follows is couched in terms of building projects only, for simplicity, but the points made apply equally to civil engineering projects.

In the UK the large majority of firms offering this type of service originated as contractors and many also offer competitive contracting as well as a design-and-build service. There is therefore a tendency for firms to be orientated towards

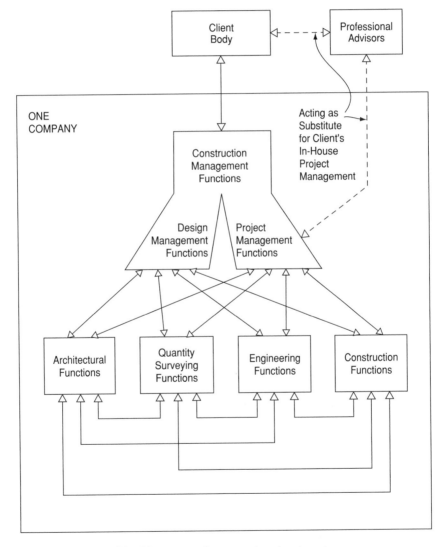

Fig. 10.4 Design-and-build structure (construction dominant).

construction activity, which may have detrimental consequences for the integration of design and a subsequent effect upon its quality. The relationships that emerge may be as shown in Fig. 10.4. For such a structure the client would need to be assured that the construction management emphasis is not allowed to dominate the project management needs of the project. A structure which is more likely to be acceptable to the client is shown in Fig. 10.5 in which it can be seen that project management is allowed to dominate, and design and construction management are integrated in an equivalent relationship.

This latter arrangement is only likely to be adopted by a construction firm which has an 'in-house' design capability that is sufficient for the project. If the

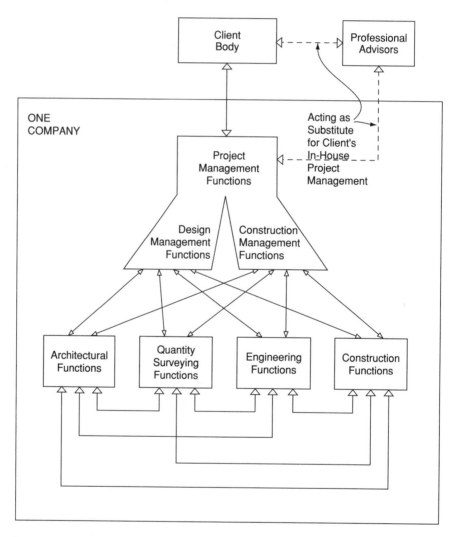

Fig. 10.5   Design-and-build structure (project management dominant).

firm has to 'subcontract' design, then the relationships that emerge are likely to be similar to a traditional arrangement in terms of the difficulty of integration. Contractors generally prefer to subcontract design (Walker 1995) due to the greater access to a wider range of design skills provided by this approach and the reduction in risk associated with not having designers on their payroll. However, the client will still retain the advantage of having only one company responsible contractually for the whole project. Similarly, the responsibility for subcontractors for specialist construction will be totally with the design-and-build firm and integration will not normally be complicated by nominated subcontractor relationships.

Design-and-build may potentially provide the most effective integration but there remains difficulty in integrating the project team and the client. The client needs to protect its position so that the project it receives on completion fulfils its requirements. The client should have a clear conception of its objectives, but those of the design-and-build firm may at times conflict with those of the client.

If, for instance, a problem needs resolving, in which the achievement of the best design solution conflicts with the method of construction, the design-and-build firm may seek to solve the problem by opting for ease of construction at the expense of the best design solution. Other similar problems may occur, for instance economy versus form of construction, speed versus construction method, etc. The client will be in a position to resolve situations to its benefit if it has sufficient in-house expertise to understand the issues and the appropriate contract conditions that allow the client to act to produce a result to its benefit. If the client has not, then professional advice upon which to act will be needed.

Professional advisors in this capacity would act as a substitute for the client's in-house project management team as shown in Figs 10.4 and 10.5. Integration between the professional advisors and the management structure of the design-and-build firm would have to be carefully designed to ensure that it had effective lines of communication and authority, which would have to be made explicit in the contract conditions, e.g. whether they are in an advisory or executive position. Naturally, this implies that the professional advisors should be closely involved in drafting the conditions of contract and in establishing the project.

An extension of the design-and-build approach that has gained more acceptance in recent years in the UK and is more widely used abroad is that, rather than arranging a contract with one selected contractor, competition takes place for both design and price. Alternatively, an upper limit on the price may be fixed so that essentially competition is based primarily on design. The management arrangements with the successful bidder would be identical to those discussed earlier for one selected firm but either the client's in-house team or professional advisors would need to evaluate the submissions against the client's criteria to advise on the bid to be accepted. In such an arrangement, bidders would be provided with details of the client's requirements, which would normally include a performance specification. These details would form the basis of the criteria against which bids would be judged. Therefore integration of the professional advisors with the client or integration within the client's organisation with its own in-house team is of paramount importance in drafting the client's requirements for the bidders and in evaluating the bids.

Prime contracting

Prime contracting can be seen as an extension of design and build (Chartered Institute of Building 2002), but on a grand scale. It originated in the 1990s as the model for procurement of construction and maintenance services for the large estates of the Ministry of Defence. It is now seen as an appropriate method for organising construction and maintenance work for any client with a large high-value ongoing demand for these services.

Prime contracting entails the appointment of one company (the prime contractor) to manage this process, with payment linked to performance and innovation. A main objective is to enable the prime contractor to invest in approaches which increase effectiveness such as supply chain management, standardisation, bulk purchasing, enhanced co-ordination and through-life costing.

Such an arrangement allows the client to work together with the prime contractor on long-term planning of the estate so that scheduling of work and enhancements can be arranged to the client's advantage. This process contrasts with a situation where contracts are let piecemeal as and when required. Even if the client has an in-house construction administrative capability, there are potential benefits from being able to closely involve a major contractor/facilities provider in the process as a 'permanent' part of organisation. Prime contracting can be seen as an outsourcing mechanism which has the potential to provide expert advice and procurement capabilities. A major contribution to the achievement of this potential will be the opportunity for greater integration of the process of managing the estate. There will be three major nodes of integration: within the client organisation, within the prime contractor organisation and between the client and prime contractor organisation, with the latter being particularly significant.

An illustration of a transaction cost explanation

The relationship between project management theory and transaction cost economics can be illustrated by a comparison of a traditional organisation structure, with the architect as designer and project manager and the contractor appointed in competition, and a design-and-build structure. This is achieved by analysing why the design process is separated from the construction processes in the traditional system. Textbook explanations in favour of the traditional system include tradition, professionalism, division of labour and flexibility. The problems created are also well documented; lack of buildability, communication problems, co-ordination problems and unclear responsibility between designer and contractors, to name a few. However, whatever arrangement is chosen is done so voluntarily by all parties. The obvious question is: if there exists a better (more efficient) alternative arrangement that consumes less resources, why didn't the parties adopt it and share the benefits?

Transaction cost economics argue that the choice of a structure is a result of the relative costs of specifying the nature of the project arising from different structures and depending on the type of project. The costs of managing the process of project specification and preparing the actual detailed specification are transaction costs. The management activities required to achieve this are defined by project management theory.

The distribution of project management activities varies considerably between traditional and design-and-build structures. In the former the client organisation will be involved in integrating with the project team in order to transmit their ideas, the architect will undertake a large number of the project management functions but other consultants will also be involved in some project management work, as will the contractor. Thus transaction costs will be distributed widely. In the case of design-and-build, the client will still be involved in project management but a large part will be accounted for by the design-and-build company. If the client also uses separate consultants to monitor the design-and-build company then their costs would also be transaction costs.

To explain this further, imagine a situation where the design-and-build structure is the only procurement system that is allowed: how could the willing buyer (client) and the willing seller (the design-and-build company) meet and agree at a price for the project? How could the client ensure that the product delivered by the contractor was the project that the client wanted to purchase in the first place? These problems cannot be resolved without a clear description of the project being traded.

The project is only specified in detail when its design is complete. When the design is not complete, the project can only be described by defining the client's requirements. Such requirements are essential for the design-and-build contractor to quote and agree a price with the client. For most buildings, the costs of defining requirements in detail would be very high as the client's requirements are often in very vague terms and perception orientated (for example aesthetic requirements). The nature of such descriptions, even when defined in a detailed manner, is often not sufficient to determine the physical characteristics of the project. Such difficulties can be overcome if a designer is employed by the client to design the project before the constructor is chosen (the traditional system). Once the project is designed, it would be much easier to specify and price the project. Although such an arrangement is costly, in most cases it is still cheaper than the design-and-build arrangement. However, if the client's requirement can be defined clearly and (relatively) cheaply, for example when the requirements are simple and more functional than perception orientated (such as low rise industrial buildings, public housing and many civil engineering works), design-and-build contracts would be adopted.

In practice this theoretical scenario may not be implemented due to a client's risk profile. In cases in which the above analysis would result in the choice of design-and-build, a client who is risk averse may perceive a greater risk in such an approach if they fear that the definition of its requirements may not be undertaken effectively, and as a result may choose a traditional approach. The client may perceive that it can better protect itself contractually using the traditional approach but this would incur transaction costs. The perceived risk under a design-and-build structure could theoretically be removed by a greater and greater level of detailed definition of requirements but this would of course, generate further transaction costs. So ultimately risk can be seen and analysed in transaction cost terms. However, this theoretical scenario does not prevent clients making suboptimal choices if the parties do not adopt transaction cost minimisation.

Organisation matrix

At their simplest level, organisation structures of projects can be seen to consist of three major components – the client, the design team and the contractor. The client's experience of construction, the organisation of the design team and the method of appointment of the contractor will have a fundamental influence on the effectiveness of the project organisation. Examples of the range are:

(a) *Client*
   (i) No construction expertise, a senior manager liaises between client and project team.
   (ii) In-house expertise available, project manager appointed within client organisation.

(b) *Design team*
   (i) Traditional organisation
   (ii) Non-executive project manager.
   (iii) Executive project manager.

(c) *Contractor's appointment*
   (i) Selective competitive tender.
   (ii) Two-stage competitive tender.
   (iii) Competitive serial tender.
   (iv) Negotiated tender.
   (v) Management contract.
   (vi) Separate trade contracts.
   (vii) Design-and-build (overlaps with (b) above).

   These classes are not exhaustive, but they represent the more common classifications and themselves produce a 2 × 3 × 7 matrix, giving 42 alternative arrangements.
   Each arrangement will present certain advantages and disadvantages and should be selected for use in circumstances that suit the particular project. The features of each arrangement are as summarised briefly in Table 10.1 and whilst they are drafted in terms of building projects, the structures are also relevant to civil engineering.

Table 10.1 Matrix of project organisation structure.

Client	Design team	Contractor's appointment
(1) No construction expertise	Traditional	Selective competition

Comments
This is the traditional arrangement and relies upon the architect as designer and manager, with the client having a limited management contribution to its project. The contractor is not integrated into the design phase. Suitable for relatively simple projects in terms of both complexity of construction and environment for which client's requirements are clear.

(2) No construction expertise	Traditional	Two-stage, competitive

Comments
Opportunity to use contractor's expertise during part of design phase depending upon when first-stage tenders are initiated. Gives opportunity to speed up programme by over-lapping construction and some design work. Appears to be a half-hearted attempt to integrate the contractor.

(3) No construction expertise	Traditional	Competitive serial

Comments
As (2) above but used where a number of similar buildings are required for the same client.

(4) No construction expertise	Traditional	Negotiated

Comments
This arrangement enables the contractor to be integrated at a very early stage in the project. Relies upon the architect as designer/manager being prepared to use and trust the contractor's expertise. Suitable for complex projects and/or environments.

(5) No construction expertise	Traditional	Management contractor

Comments
In management terms, similar to (4) above but allows for competition for construction work. Still relies upon the architect as designer/manager being prepared to use and trust the contractor's expertise. The management contractor is acting as another consultant with responsibility for arranging and organising subcontractors and may be more acceptable to other members of the design team in this capacity. Potential conflicts between architect acting in a management capacity as well as being designer and the management role of the management contractor. Suitable for complex projects and/or environments. Illustrated in Fig. 10.6.

(6) No construction expertise	Traditional	Separate trade contracts

Comments
Potentially high in integration by direct contact between the design team, particularly the architect and subcontractors on site which could be extended into the design phase. Re-duces differentiation of major management components by eliminating a separate general contractor function. Requires an architect leader who is strongly management orientated and objective about the often conflicting demands of design and construction. Illustrated in Fig. 10.7.

(7) No construction expertise	Traditional	Design-and-build

Comments
This arrangement would not be used as described. The conventional design team would act as consultants to the client to monitor the design-and-build contract. As the client has no in-house expertise, some professional advice would be required to assist the client's representatives in dealing with the design-and-build contractor. The conventional design team in this context would probably consist of one or two professional advisors (or a firm). Integration should be strong within the design-and-build firm but the potential

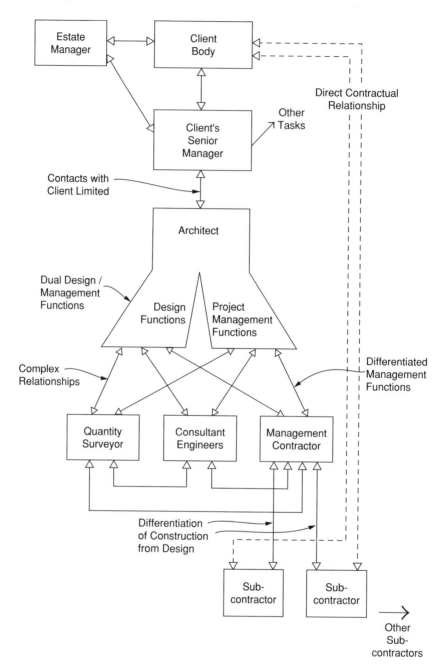

Fig. 10.6   Diagram of type 5 structure.

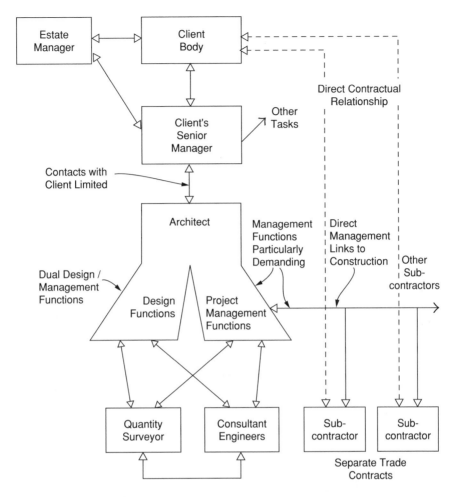

Fig. 10.7 Diagram of type 6 structure.

Table 10.1 (*Continued*)

Client	Design team	Contractor's appointment

weakness is in the integration with the client when it has no in-house expertise, even though it has appointed advisors. Illustrated in Figs 10.4 and 10.5.

(8) In-house expertise	Traditional	Selective competition
(9) In-house expertise	Traditional	Two-stage competition
(10) In-house expertise	Traditional	Competitive serial
(11) In-house expertise	Traditional	Negotiated
(12) In-house expertise	Traditional	Management contractor
(13) In-house expertise	Traditional	Separate trade contracts
(14) In-house expertise	Traditional	Design-and-build

Table 10.1   (*Continued*)

Client	Design team	Contractor's appointment

Comments

This group is similar to (1) to (7) above except that the client has in-house building expertise available and appoints a project manager from within its own organisation.

The result in each case is that integration between the project team and the client should be closer, provided that the client's own internal integration is effective. A member of the client's organisation appointed project manager should have more time to devote to the project than if a senior manager were doing this job in addition to normal work. The demands on the architect to exercise this management role effectively are likely to be greater as a result of the pressure exerted by the project manager.

Approaches to appointing the contractor and integrating the contractor into the team other than by selective competitive tender are more likely to be adopted because of the influence of the client's in-house project manager.

Alternative (14) is unlikely to be used as the in-house project manager should normally have the expertise and capability to monitor the design-and-build contract. Alternative (13) is illustrated as an example in Fig. 10.8.

Client	Design team	Contractor's appointment
(15) No construction expertise	Non-executive project manager	Selective competition
(16) No construction expertise	Non-executive project manager	Two-stage competition
(17) No construction expertise	Non-executive project manager	Competitive serial
(18) No construction expertise	Non-executive project manager	Negotiated
(19) No construction expertise	Non-executive project manager	Management contractor
(20) No construction expertise	Non-executive project manager	Separate trade contracts
(21) No construction expertise	Non-executive project manager	Design-and-build
(22) In-house expertise	Non-executive project manager	Selective competition
(23) In-house expertise	Non-executive project manager	Two-stage competition
(24) In-house expertise	Non-executive project manager	Competitive serial
(25) In-house expertise	Non-executive project manager	Negotiated
(26) In-house expertise	Non-executive project manager	Management contractor
(27) In-house expertise	Non-executive project manager	Separate trade contracts
(28) In-house expertise	Non-executive project manager	Design-and-build

Comments

This group corresponds with (1) to (14) above, except that a non-executive project manager is appointed within the project team. As discussed earlier, someone in this position will fill a co-ordination and communication role without authority for executive functions. Provided that the role is recognised and accepted by the other team members, the administration of the project should benefit. But as the decision-making structure and authority pattern remain unaltered, the effect on the management of the project is unlikely to be significant and the comments given for (1) to (14) will apply. The non-executive project manager is likely to be dominated by the designer and in the cases where the client appoints an in-house project manager, he or she will be particularly easily overridden. Alternative (25) is illustrated in Fig. 10.9.

Client	Design team	Contractor's appointment
(29) No construction expertise	Executive project manager	Selective competition

Comments

Management and design responsibilities are split, which should allow the project manager to concentrate upon management of the project both within and between the design and construction teams and with the client. Probably the most effective way of improving the management of what is still really a conventional structure, suitable for complex projects and/or environments, where it is necessary for the contract to be awarded competitively.

Client	Design team	Contractor's appointment
(30) No construction expertise	Executive project manager	Two-stage competition

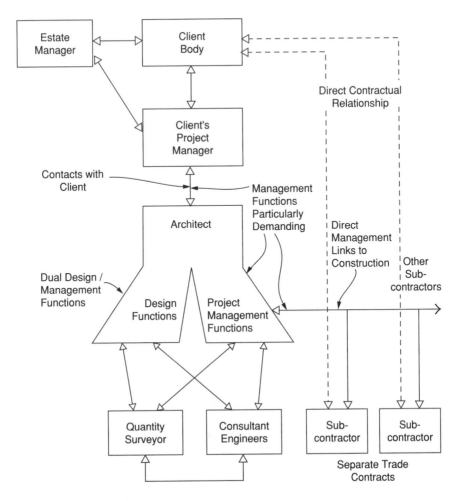

Fig. 10.8 Diagram of type 13 structure.

Table 10.1 (*Continued*)

Client	Design team	Contractor's appointment

Comments
Comments as for (2). Opportunity is presented for the project manager to ensure that the contractor is properly integrated and makes a contribution to the design phase. It will be up to the project manager to time the first phase so that the contractor's contribution is maximised. Illustrated in Fig. 10.10.

(31) No construction expertise   Executive project manager   Competitive serial

Comments
As (30) above, but where a number of buildings are required for the same client.

(32) No construction expertise   Executive project manager   Negotiated

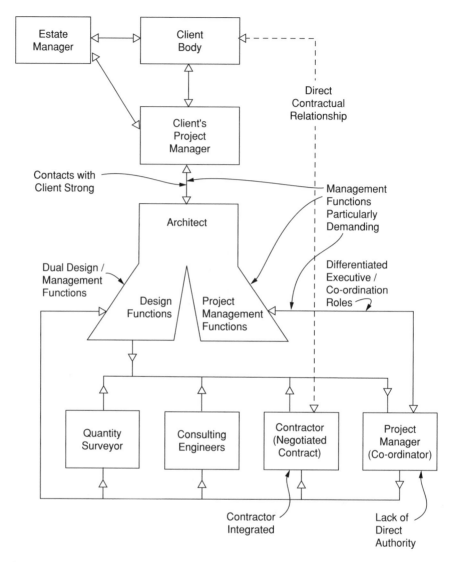

Fig. 10.9   Diagram of type 25 structure.

Table 10.1   (*Continued*)

Client	Design team	Contractor's appointment

Comments
This arrangement allows the project manager to establish a tightly integrated team from the very early stages of the project. Allows the project manager to appraise the contribution from all members of the project team objectively. Suitable for very complex projects and/or environments.

(33) No construction expertise   Executive project manager   Management contractor

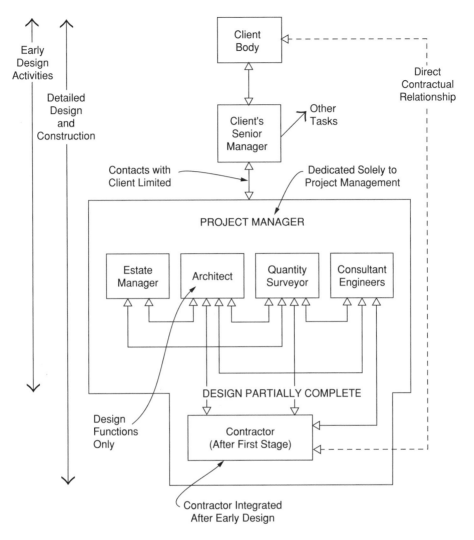

Fig. 10.10 Diagram of type 30 structure.

Table 10.1 (*Continued*)

Client	Design team	Contractor's appointment

Comments
In management terms similar to (32) above, but allows for competition for construction work. As the management contractor is acting as another consultant, the project manager may be able to integrate the management contractor with less constraint than may be the case with (32). Suitable for very complex projects and/or environments. Illustrated in Fig. 10.11.

(34) No construction expertise	Executive project manager	Separate trade contracts

Fig. 10.11   Diagram of type 33 structure.

Table 10.1   (*Continued*)

Client	Design team	Contractor's appointment

Comments
Potentially high in integration by direct contact between the design team, particularly the project manager and the subcontractors. Reduces differentiation of major management components by eliminating a separate general or management contractor function. The use of a project manager who is strongly management orientated and objective about the often conflicting demands of design and construction should provide the foundations for a successful project. Suitable for complex projects and/or environments.

Table 10.1   (*Continued*)

Client	Design team	Contractor's appointment
(35) No construction expertise	Executive project manager	Design-and-build

Comments
In this arrangement the executive project manager would not have a supporting design team but would act as the client's representative in monitoring and controlling the design-and-build contract. Integration should be strong within the design-and-build firm. Integration between the client and project manager and between the project manager and the design-and-build firm would depend to a large extent on the relationships established in the formal contract. The project manager should be involved in establishing the contract in which his or her authority should be clearly established.

Client	Design team	Contractor's appointment
(36) In-house expertise	Executive project manager	Selective competition
(37) In-house expertise	Executive project manager	Two-stage competition
(38) In-house expertise	Executive project manager	Competitive serial
(39) In-house expertise	Executive project manager	Negotiated
(40) In-house expertise	Executive project manager	Management contractor
(41) In-house expertise	Executive project manager	Separate trade contracts
(42) In-house expertise	Executive project manager	Design-and-build

Comments
This group is similar to (28) to (35) above, except that the client has in-house building expertise available and appoints a project manager from within its own organisation in addition to the executive project manager of the project team. The result in each case is that integration between the project team and the client should be closer, provided that the client's own internal integration is effective. A member of the client's organisation appointed project manager should have more time to devote to the project than a senior manager who may also have to do routine work. The use of the two project managers should strengthen the integration with the client to the extent that this group of arrangements should represent potentially the strongest management-dominated organisational structure. Alternative (41) is illustrated in Fig. 10.12. In particular the use of structures that allow the contractor to be integrated into the team (39, 40, 41) has the potential for full integration and allegiance to objectives of even the most complex projects, the facility for implementing the most rigorous control mechanisms and the opportunity to take the most appropriate advice before decisions are taken.

Whilst it has been recognised that the selection of procurement systems by clients has become increasingly complex (Masterman & Gameson 1994), one of the key issues in the choice of organisation structure lies in the trade-off between the apparent competitiveness of the bid price for the construction work and the early involvement of the contractor in the project team. Although competition for a construction contract may appear to provide the client with the lowest price, it is achieved only at the cost of a potentially less integrated project team, as the contractor cannot be brought into the design team sufficiently early to influence the 'buildability' of the design. The effect of this could be a longer construction period and a higher price because of the difficulty of construction of the proposed design. Hence the competitiveness of the bid may be more imagined than real. In the interests of the total economy of the project, a fully integrated team throughout the whole process may be more beneficial to the client's interests. This

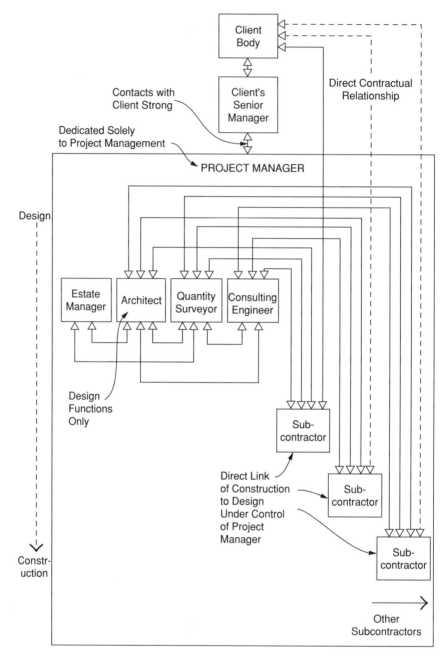

Fig. 10.12   Diagram of type 41 structure.

factor will be most important in selecting the appropriate structure of the major components of the project organisation to suit the particular circumstances of the project and the needs of the client.

Whilst the need to design the organisation to suit the particular circumstances of the project is generally accepted, the search for the Holy Grail continues. An example is the idea of technology clusters as described by Gray (2004). He says: 'the objective is to merge together the work group with the support function into an integrated task group called a technology cluster' and 'a technology cluster must be initiated at the earliest possible point, preferably before design commences... The aims of technology clusters are to:

- 'Group all contributors together, preferably in one location.
- Elicit solutions to technical, quality and efficiency criteria to support innovative design solutions.
- Create a fully integrated systems-level solution.
- Focus on completion of the system as an integrated unit.
- Preserve the value chain throughout the supply chain.'

Upon application to two projects, it was found that lessons could be learned. These were that 'world class' companies were needed, trust and transparency were difficult to achieve and the role of the design team was ambiguous; but that improved collaborative working was a major achievement.

Whilst the search for new organisation structures for construction projects will continue, it is interesting to note that ideas underpinning technology clusters continue to reflect the systems approach. The deficiencies of the arrangement appear to be associated with two factors other than systems applications and one associated with systems design (i.e. integration of the design team).

Public–Private Partnerships

Put simply, a public–private partnership (PPP) is an arrangement between a public sector organisation and a private sector organisation for the provision of a public sector facility e.g. hospital, school, road, which will be provided, owned and operated by the private organisation for a specified period before reverting to the public body. During this period the public body will pay the private body for the use of the facility. There are many variations of this relatively simple type described here, but all will contain many of these elements. Whilst the term PPP may still appear rather new, the concept has been around for many years in the guise of build-operate-transfer (BOT) or build-own-operate (BOO). In the UK, the use of PPP projects has been stimulated by the government's private finance initiative (PFI), which also incorporates the provision of finance by the private body in forms such as design-build-finance-operate (DBFO) and its variants.

The idea is that both sectors bring their skills to the project which, hence, gains from such synergy; the DBFO form affords the greatest potential for such gains. Countering such potential is the increase in risk to the private body consequent on the greater complexity of the organisational and financial structure associated

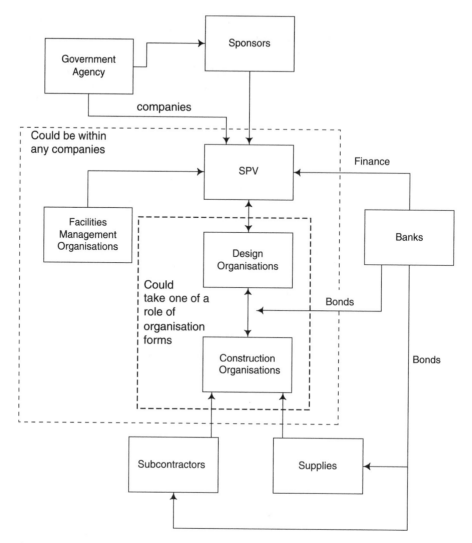

Fig. 10.13   Indicative arrangements for a PFI scheme.

with this arrangement and the risk of an unsuccessful project outcome for the public body, although it has been claimed that risk is in reality not transferred to the private sector to the extent expected by the public body (Fryer *et al.* 2004).

The aspects of PPP/PFI of concern to this book are the organisational issues arising from the structure's use (Fig. 10.13). These issues relate to the relationship between the public body, usually termed the 'sponsor', and the private body, often formed specifically for the project, called the 'special purpose vehicle' (SPV), as well as to the actual composition of the SPV itself, which may be a consortium of private entities. The relationship between the sponsor and the SPV is akin to the relationship of the client and contractor on conventional construction projects, but much more complex. Hence, differentiation and integration remain of major concern. Differentiation is likely to be great as, although it is claimed that PPP/PFI creates converging objectives between the parties, the mindsets of the public sector and private sector people are likely to diverge, creating a need for strong integration which in turn will require project management of a high order.

The nature and hence the structure of the SPV itself can take many forms. In its simplest form it could be an organisation which has a design, construction and operational management capability which does not need to hire outside organisations to carry out many tasks; this would be very rare, although some companies could be set up in this way, in which case differentiation would be kept to a minimum. At the other extreme, the SPV could be specially set up for a particular PFI project and may hire all its capability (design, construction, etc.) except management, in which case differentiation would be high and require great integration effort. Between these two extremes there are a myriad of alternative arrangements, each generating its own integration needs and each of which would be heavily influenced by the construction procurement method. An indication of the variety of arrangements is given by illustrations in the earlier section 'Organisation Matrix', but for PFI projects, which are invariably complex, a number of procurement methods may be used for different parts of the project. Similarly, a range of specialist designers are likely to be used, leading to even greater differentiation.

A major objective of PPP/PFI is the allocation to the private sector of those risks which are best managed by it – such as design, construction and operation risks – and the retention by the public sector partner of risks better managed by it (Li & Akintoye 2003). The risks are extensive; Hardcastle and Boothroyd (2003) list 40, one of which is organisation, defined as 'the resource availability and capability of the performing organisations'. In this context, the performance of the overarching project management system is critical for project success. Not only are PPP/PFI projects amongst the most complex from a management perspective, the differing objectives between the public body and the private entity and the public's expectations can lead to difficult relationships, which has meant that PPP/PFI projects have attracted criticisms (Li *et al.* 2005; Fryer *et al.* 2004). Of 18 critical success factors identified by Li *et al.* (2005), three concerned organisation: good governance, well organised and committed public agency, and shared authority between the public and private sectors. It should be added that the survey from which they were identified had more private sector respondents than public sector, and that the true success of a project should be measured using inputs from users, which is not yet possible for many PPP/PFI projects as they have not been in use for long.

The factor that militates against achieving best value in PFI project procurement is, essentially, high transaction costs, which include lengthy and complex negotiations, difficulty in specifying requirements, and difficulty in pricing (Akintoye *et al.* 2003), all of which reflect Dahlman's (1979) classification of transaction costs as search and information costs, bargaining and decision costs, and policing and enforcement costs. High transaction costs are coupled with the public sector's lack of ability to manage consultants.

Programme management

Programme management has been recognised only recently as something that should be of interest to managers of construction projects, although it has been recognised for longer in other industries. It has come to prominence in construction through the suggestion that its ideas should be relevant to programmes of construction projects (Blismas *et al.* 2004). Programme management as a homogeneous concept is difficult to pin down (Partington *et al.* 2005). Blismas *et al.* refer to interchangeable terms relating to programme management as portfolio, multi-project environment and programme. They define portfolio as 'a collective group of construction projects and programmes within an organisation, with no interference to the manner in which they are organised or managed'. Multi-project environment is 'used to emphasize the multiplicity of projects in an organisation'. Programmes are 'groupings of projects within the broader portfolio' and a 'framework that consists of multiple interdependent projects, is long-term or indefinite, focuses on the benefits or strategic aims of an organisation, provides common purpose between projects and is usually a large undertaking'. They then point out that a survey of definitions 'reveals a diffusion of defining characteristics'. To say that the definition of programme management is hazy would be something of an understatement.

Papers on programme management and comparable activities seem to be at great pains to state that it is not the same as project management and that effective project managers do not necessarily make effective programme managers (Blismas *et al.* 2004; Partington *et al.* 2005; Lycett *et al.* 2004). The point frequently made is that single-project management is not transferable to multi-projects. It is surprising that such an argument needs to be made in the literature, as the nature of project management and programme management is fundamentally different. Programme management is essentially a business strategy that is project-driven (Fewings 2005) and thereby operates at a different level within organisations. There is an obvious difference in the talents needed to manage a single project than those required to manage complex, shifting, strategic programmes (Partington *et al.* 2005). Presumably firms assumed that project managers could undertake programme management as they appeared to offer the nearest comparable skills, without realising the different expertise needed for strategic and tactical management.

Whilst different skills may be needed for project and programme management, nevertheless programme management can be seen to be part of the *project management process* of organisations that have streams of projects. It is the overarching strategic management activity; there is 'an increasing recognition that programme management provides a means to bridge the gap between project delivery and organisational strategy' (Lycett *et al.* 2004). Research on programme management has identified three types of programmes: bounded programmes which are well defined, of limited scope and, hence, have high levels of certainty; rolling programmes, which are the opposite of bounded programmes as they have a loose, ongoing nature and hence uncertainty; and target programmes, which are a hybrid form of the previous two types. It is suggested that 'bounded and some target programmes could be procured more efficiently as entire pro-grammes rather than individual projects, whereas rolling programmes would rely on groups of preferred suppliers negotiating projects with the client as they occurred' (Blismas *et al.* 2004).

Definition and development of programme management skills is still in its in-fancy and seems to be concerned more with what programme management is not, rather than with what it is. However, Lycett *et al.* (2004) give helpful guid-ance in identifying that the problems experienced when programme management techniques are applied in practice can be related to the ineffective management of one or more of:

- The relationship between the programme manager and the project managers within the project
- The relationship between the constituent projects of the programme and the wider business context
- The relationship between the individual project managers within the pro-gramme.

Interestingly, therefore, on the basis of these problems, systems theory would seem to have much to offer to their understanding.

Chapter 11
Analysis and Design of Project Management Structures

Need for analysis and design

The mainly theoretical scenario developed so far provides the concepts necessary for analysing project organisations and the basis for a structured approach to their design. Such theory is all very well but it needs translating into techniques that are useful in practice and make a positive contribution to improving the effectiveness of the management of projects in the real world. For analysis, such techniques should be capable of mapping what actually happened on projects. When used for design they should be capable of showing clearly what is expected to happen if the project is to be managed effectively. The predominant need is to design structures that allow people to work together effectively, but there is also the parallel need to develop structures that enable the use of appropriate project management techniques.

The increasingly explicit recognition of project management in its own right rather than as something subservient to other professional skills has helped to generate a range of techniques and tools for project control, e.g. critical path networks, line of balance, etc. The opportunity and the will to employ such techniques depend upon a receptive management structure and the effective organisation of contributors. A management structure led by people whose priority is management should result in more widespread use of project control techniques and ensure that contributors to the project are organised to maximise the benefits of such techniques. But what is often overlooked is that it is the responsibility of those exercising project management skills on behalf of clients to design organisation structures appropriate to particular projects and their environments so that the right skills and techniques are used at the right time. This is particularly difficult to achieve in construction as the structures normally used are predominantly conventional and reflect the juxtaposition of traditional professional roles. This tends to inhibit innovation, with the result that the industry and professions are slow to apply new ideas and techniques.

Criteria

Techniques for analysing and designing organisational structures should make clear the following aspects of the way in which the project is organised:

- the operating system
- the managing system
- the relationship of people in the organisation and their interdependency
- the roles of the people in the organisation
- the position of the decision points and their status, e.g. key, operational
- the contribution of people to each decision and their relationships in arriving at decisions.

An approach that exposes these aspects will give great visibility to the way in which projects are organised. It will show clearly who did what and, perhaps more significantly, who did not. Compared with the traditional pyramidal organisation chart, such an approach would, if used for organisation design, be a dynamic representation of what should happen on a project, or, if used for analysis, what really did happen, rather than being a simplistic, static statement of who is whose boss, without any attempt to relate the people to the project activities and to each other.

Linear responsibility analysis and other techniques

Three techniques are useful in satisfying the criteria: *transformed relationships evolved from network data* (TREND), *linear responsibility analysis* (LRA) and *social network theory* (SNA).

TREND (von Seifers 1972; Bennigson & Balthasas 1974) was developed to analyse organisations in a study of temporary management structures. It has the potential to provide a powerful and sophisticated tool for project organisation design where a high level of uncertainty needs to be managed. The categories of information required were identified as:

- What tasks are to be performed?
- Who is responsible for each task?
- How do the tasks interrelate?
- What is the interaction pattern of the participating departments?
- Which tasks and departments are more critical to successful completion?
- What is the nature and location of the uncertainties involved?

It was recognised that much of this information could be obtained from a network, but that to answer the last two questions a PERT (Programme Evaluation and Review Technique) network with three time estimates for activities would be required. However, very few projects are available with networks to such a level of detail.

An alternative approach, which does not require extensive networks to be available and which incorporates within it a technique for building up organisation networks at various levels of detail, has been developed from linear responsibility charting. This first appeared in the mid 1950s (Anon 1955) but does not appear to have been further developed until about 1975, by Cleland and King (1983). Linear responsibility charting originated as an improvement upon the pyramidal organisation chart so that it shows who participates and to what degree when an activity is performed or a decision made.

It was subsequently discovered that LRC could serve as a tool for organisation design and analysis since it can be made to display system interfaces and interrelationships. It was then further advanced into the linear responsibility analysis (LRA) technique in 1980 (Walker 1980) and has subsequently been used for a number of studies of project management, of which a comparison of two editions of the RIBA *Plan of Work* (Hughes 2003) is a recent example. LRA allows the degree and quality of differentiation, integration and interdependency within an organisation system to be identified and was applied to building projects (Walker 1980). Decision points were overlain on the LRA to show the combination of contributors to decisions and their roles.

Although arising from general management thinking, LRA is particularly useful for project management. Its effectiveness lies in its ability to expose both the decision points in the process and the way in which the people in the operating and managing systems are arranged in relation to those decision points. It also allows identification of the activities of the people in the managing and operating systems and of the relationships between them. Thus LRA gives visibility to the arrangement and integration of contributors bringing forward propositions upon which decisions are based. An understanding of how this process works is fundamental to the success of a project organisation and LRA makes a significant contribution.

Whilst LRA works well at the level of the individual project as an aid to effective organisation and timely decision making, social network theory (SNA) is proposed for the comparative analysis of procurement and project management arrangements. In particular, it 'enables traditional project coalition management approaches to be compared with those associated with innovative management approaches' (Pryke 2004, 2005). Pryke states that SNA is based on graph theory (Scott 1992) and represents organisational groupings as systems of nodes or *actors*, joined in permanent or (as in the case of construction coalitions) transitory configurations. These networks of nodes (which, for our purposes, are firms) are linked by relational ties, which can take a number of forms (financial, legal, friendship, for example). SNA 'generates notoriously high volumes of data' which are collected using 'a form of linear responsibility analysis chart' (Pryke 2004, 2005).

Application of linear responsibility analysis

Starting with the linear responsibility chart (LRC) from which the linear responsibility analysis (LRA) originated, the best way to understand the approach and how information is presented is to examine a typical chart such as that for the design of a building project, shown in Fig. 11.1.

The LRC illustrated uses eleven symbols (see page 279 for detailed description) to indicate different types of relationship that may exist between any job position in the organisation structure and any task to be performed. The job positions are listed along the horizontal axis of the matrix and the tasks to be carried out

Legend:

- ○ Did The Work
- ◀ Approves
- ▶ Recommends
- ● General Oversight
- ◆ Direct Oversight
- △ Boundary Control
- □ Monitoring
- ◇ Maintenance
- ■ Consultation - Gave Instructions And Information
- ▽ Consultation - Gave Advice And Information
- ⊗ Output Notification Mandatory

MAJOR TASK

1. Identify Need for Project
2. Define Outline Requirements
3. Establish Budget Estimate
4. Presentation for Inclusion in 5 Year Plan
5. Programme Proposals
6. Contractual Proposals
7. Spatial Proposals
8. Technical Proposals (structural)
9. Technical Proposals (Services)
10. Technical Proposals (Architectural)
11. Financial Proposals
12. Consolidate Brief
13. Capital Expenditure Presentation
14. Programme Details
15. Working Drawings
16. Technical Details (Structural)
17. Technical Details (Services)
18. Technical Details (Architectural)
19. Contract Details
20. Contract Documentation

Fig. 11.1 Typical linear responsibility chart.

are listed down the vertical axis. In the square where a job position and a task meet, the relationship is indicated by inserting the appropriate symbol. If a job position has no relationship with a task, the square is, of course, left blank.

For example, for Task 6, 'Contractual Proposals', the partner of the quantity surveying practice prepared the proposals and in doing so consulted the partner of the architectural practice, the managers of the services engineering and structural engineering departments and the senior partner of the engineering practice. This was done under the management of the project manager who was concerned with boundary control, and with maintenance and monitoring. The project manager finally recommended the proposals to the client's project engineer. The latter approved them as well as being consulted for instructions and advice during the work and exercising general oversight.

A great benefit of the LRC is the virtue of presenting much in little space in a dynamic form, but it is much more than this. It gives an overall perspective of a project organisation structure which brings to life relationships in a way which more conventional approaches such as pyramidal organisation charts do not. It also gives a basis for further development of more sophisticated and effective techniques of organisation design.

The degree of detail of the tasks selected for the vertical axis is under the control of the designers of the organisation, either as a result of the information which they currently have available or as a result of the particular aspect of organisation design with which they are concerned. Similarly, the job positions can be in the range from individual people to whole departments or firms. The symbols likewise can be selected for the particular purpose. The inherent flexibility of the chart is very valuable to designers in that it allows them to orientate their work to the particular problem they wish to study.

Cleland and King (1983) enhanced the LRC by visualising it as an input–output device, as shown in Fig. 11.2. The input is the person in the job position with the 'does the work' relationship to the task, and the output is the completion of the task. The inputs–outputs (or 'does the work' relationships) are then transformed into schematic form, as shown in Fig. 11.3. The connected boxes containing the job position which 'do the work' and the tasks they carry out form the operating system through which the project is achieved. To this schematic are added the job positions of those who are involved in managing the process together with the symbols representing their specific management functions in connection with each task. These are placed in loops above the task boxes and are known as the control loops. The control loops represent the managing system of the project. People in relationships other than in the operating or managing systems, mainly in consulting roles, are then added as shown. When completed for the whole project, the schematic LRC shows the way in which the tasks are connected and how people act and interact within the organisation in carrying out the project.

This presentation clarifies the operating system (the linked tasks) through which the project is carried out, and the managing system (the job positions in the control loops) which controls the operating system and the relationships of others who contribute.

A further development of the LRC with the title linear responsibility analysis involves the addition of further systems information to give a sharper view of the

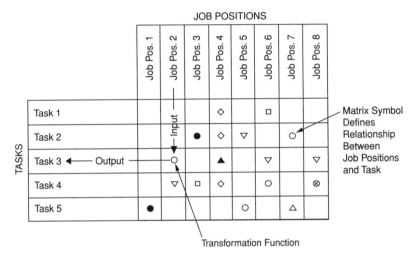

Fig. 11.2   LRC matrix (showing input–output application).

system as a whole. This total systems picture of the project enables an objective assessment to be made of the level of integration necessary for the system to work effectively. It also enables weaknesses in the system to be identified. If necessary, the system can be redesigned at those points or alternatively particular attention can be paid to integration. This development, as illustrated in Fig. 11.4, adds the following to the schematic LRC:

(a) *Interdependency:* The type of interdependency between tasks is shown, sequential interdependency by a solid line and reciprocal by a broken line. The type of interdependency influences the degree and type of integration required. Reciprocal interdependency requires greater integrative effort than sequential and needs a more flexible approach. Projects with a high proportion of reciprocal interdependencies will require great skill and effort in integration.

(b) *Differentiation:* The causes of differentiation between the contributors are shown by symbols representing the various types. As discussed in Chapter 7, differentiation is caused by:

   (i) *technology* (or skill – the technical demands of the job someone does, which determine the way in which work is divided between groups of people (T_1);

   (ii) *territory* (or location) – the geographical distance between people (T_2);

   (iii) *time* – the sequence of people's work on the project (T_3).

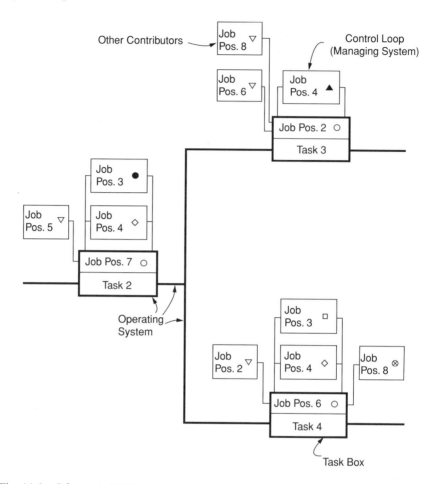

Fig. 11.3   Schematic LRC.

The factors creating differentiation can also be reinforced on construction projects by *sentience*. A sentient group is one to which individuals are prepared to commit themselves and on which they depend for emotional support. On the LRA it is identified as:

(i) sentience arising from professional allegiance (S_1);
(ii) sentience arising from both professional allegiance and allegiance to a firm (S_2).

The greatest degree of differentiation that can be shown on the LRA between contributors is therefore represented by the symbols T_1, T_2, T_3, S_2. The minimum is when there is no differentiation present in the relationship.

An examination of the degree of differentiation shown on an LRA will indicate the amount of integrative effort that will be necessary and in particular when and where on the project integrative effort is likely to be especially important. Conversely, areas of the project that should require relatively low levels of integrative effort will be revealed.

Fig. 11.4   Linear responsibility analysis in principle.

(c) *Decision points:* The decision points at the various levels of the hierarchy (primary, key and operational) are overlain on the LRA. When designing an organisation structure, the decision points have to be anticipated. This is part of the planning process and requires the close collaboration of the client. A particular skill is to ensure that the arrangement of contributors is appropriately designed prior to each particular decision. When used for analysis, the arrangement of the contributors relative to the decisions made is exposed for examination. As referred to previously, the decision points provide feedback opportunities which should be taken for the project to be completed successfully. They can be indicated on the LRA together with details of the control against which the state of the project's development has to be measured.

(d) The systems and sub-systems that make up the total project system can now be identified. They will overcome conventional professional boundaries and enable the project participants to visualise the project in terms of interrelated tasks and people rather than in terms of professional compartments.

An example of a small part of an LRA of a completed project used for post-mortem purposes is shown in Fig. 11.5.

Matrix symbols

The symbols define the way in which job positions relate to tasks and are selected and defined to suit the particular needs of the project. Each relationship can be classified into one of four categories:

 (i) a transfer function of input into output within the operating system
 (ii) a control loop function concerned with managing the operating system
(iii) a contribution of input to a task, external to the operating system
(iv) a receipt of output from a task, external to the operating system.

Examples of relationships appropriate to the construction process, derived from Cleland and King (1983) are now described.

(a) *Transfer function*
   'Does the work': This is where inputs to tasks are transformed into outputs from tasks in accordance with instructions. It is the juncture of managing and operating systems where the output is transferred under the control of the managing system to be input to the next task. This relationship appears in each task box, and the total of the task boxes defines the operating system and those involved in it. It is the relationship in which professional skills are directly applied to the project, e.g. designing, constructing, preparing documentation, etc.

(b) *Control loop functions*
   (i) *Approves:* This constitutes the final control loop function. The person in this executive relationship has the authority to approve the output of tasks for use in further tasks on the project. Normally, it is to be expected that the client will retain approval power for tasks directly affecting primary and key decisions, and that the member of the project team responsible for project management will approve those affecting operational decisions and other output.

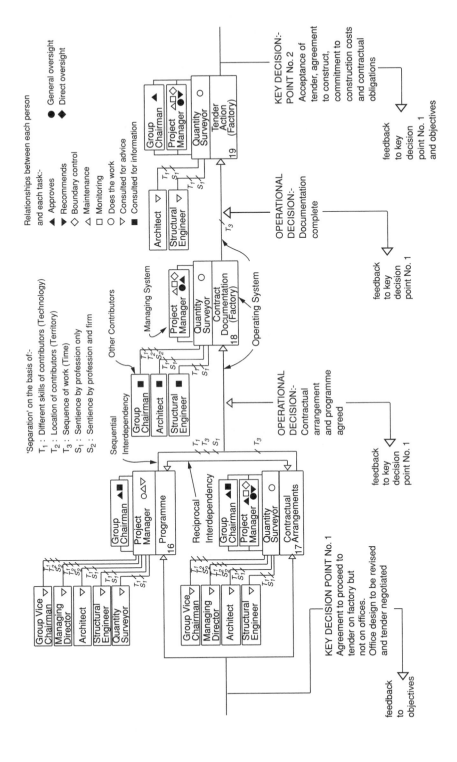

Fig. 11.5   Application of linear responsibility analysis.

(ii) *Recommends:* The person in this relationship is charged with the responsibility for making recommendations for approval of the output of tasks. The member of the project team responsible for project management will usually make recommendations to the client for approval. If the project manager is required to approve the output of the task, then some other member of the project team would normally make the recommendation.

(iii) *General oversight:* This is the broadest administrative control element and the source of policy guidance. The person in the direct oversight relationship responds to the wishes of the person in this relationship. The person in the general oversight relationship will not himself be exercising the skills of a task over which he has oversight. The primary role is to furnish policies and guidance of a scope that permits as much decision-making flexibility as possible within a task in arriving at the output. An example would be the role of a senior partner of a professional consultancy engaged to manage the project. That person would not normally be working within the operating system but would be managing the firm's input and on these occasions would appear in the control loop as part of the managing system. The omission of the relationship from a control loop indicates that the task was assumed not to involve questions of policy.

(iv) *Direct oversight:* This is the administrative control element immediately below the 'general oversight' relationship. Although having no specific project functions, the person in this relationship has, and will exercise when necessary, the skills demanded by a task over which he or she has oversight. The person is seen by others involved in the project to be maintaining a presence close to project activities. The omission of this relationship from a control loop indicates that the task was of such a routine nature that direct supervision was not necessary. An example could be a partner or senior assistant in a professional consultancy who is leading a team of people who are actually doing the work required by a particular task. The frequency of the appearance of this symbol on an LRA depends upon the level of detail of the LRA. Whereas it may not appear on an LRA prepared at a strategic level as it would be subsumed within the 'does the work' relationship in the task box, it would appear frequently if an LRA were prepared at a greater level of detail, say to plan the activities of one critical aspect of the project organisation which required close examination of a particular professional consultancy's contribution.

(v) *Boundary control:* When this appears in a control loop it indicates the specific control activity of ensuring functional compatibility within the task for which it appears and between it and other tasks. The person in this job position does not normally also have an administrative role. In addition, this relationship is concerned with relating the total system to its environment. It is a project management function as it is concerned solely with integrating the tasks and hence the people working on projects, and with ensuring that the tasks and the total system respond to changes in the project's environment. Omission of this relationship from a control loop indicates that the task is undertaken independently of other tasks in the operating system, which is an unlikely occurrence on a construction project.

(vi) *Monitoring:* This is the specific control activity of intra-task regulation concerned with checking prior to output to ensure that a 'does the work' activity is achieving its purpose. Omission of this relationship from a control loop indicates that it was not necessary to carry out such checks because there was total confidence that the activity would achieve its purpose.

Monitoring is a project management function. Although this activity may be considered to be the responsibility of the firm providing a particular contribution to the project, it is nevertheless the responsibility of project management to ensure that all aspects of the system are performing satisfactorily. Even if a particular task is receiving 'direct oversight' from a senior member of the firm providing the contribution, it is to be expected that the person providing project management would exercise a monitoring function over all activities.

(vii) *Maintenance:* This is the specific activity of ensuring that a 'does the work' activity is being maintained in an effective state, both quantitatively and qualitatively, so that it is capable of achieving its purpose. Omission of this relationship from a control loop indicates that it was not necessary to maintain the 'does the work' activity. Again, this is a project management function concerned with making sure that the right number of people of the appropriate quality are employed on a task. As with the last relationship, even though the managers directly associated with the task may also consider that this is their function, it is also the responsibility of whoever is managing the total project on behalf of the client.

(c) *Contributions to input*

(i) *Consultation – gave instructions and information:* This is an input of instructions and information to a 'does the work' activity and does not therefore appear in the control loop. Boundary control activities should ensure that the necessary people are placed in this relationship to appropriate tasks.

(ii) *Consultation – gave advice and information:* This is comparable to the last relationship but advice (rather than instructions) and information are input to a 'does the work' activity. Again, boundary control activities should ensure that the necessary people are placed in this relationship to appropriate tasks.

(d) *Receipt of output*

Output notification mandatory: This is placed in the output of a task when it is essential that the person in this relationship with a task receives timely information concerning a task output. The concept of this relationship is one of passive transmission of information.

Benefits

An LRA gives great visibility of how people work or do not work together on a project. In particular, it makes clear:

(i) *The sequence of the tasks and the effectiveness of the process in relation to the decision points and the actual decisions taken.* The tasks undertaken before a decision is made and the skills and status of the people contributing to the tasks and the decision will be clearly exposed.

(ii) *The type of interdependency, which identifies the way in which the tasks are related.* Inappropriately drawn interdependencies, e.g. sequential, which should have been reciprocal, are normally ineffective and most difficult to integrate (for instance, if the structural engineer and the quantity surveyor simply respond to the architect rather than each influencing the other's decisions).

(iii) *Who contributes to each task and with what status and what relationship to others.* More importantly, perhaps, it shows who does not contribute or who contributes in the wrong relationship. The LRA shows the degree of differentiation between contributors and hence the degree of integration required. The amount of differentiation is reflected quantitatively by the number of links between tasks and between job positions within tasks. The simpler the project the fewer links there will be. The links between tasks represent the differentiation that has to be integrated in managing the output of the tasks to realise the project. The links between job positions within tasks give the differentiation to be integrated in achieving the output of each task.

A qualitative measure of differentiation is given by identifying the proportion of each permutation of the factors of differentiation (technology, territory, time and sentience). Theoretically, it is possible to have any permutation of differentiation factors (T_1, T_2, T_3, S_1, S_2), but in practice the configuration of the contributors limits the range which occurs on any project. For example, within a task the various professions in a multidisciplinary practice in which all members are located in one office and can only have a differentiation of T_1, S_1 (technology and sentience by profession).

It is to be expected that maximum integration will only be possible if an appropriate pattern of consultation has been established between the contributors. Such a pattern of consultation is demonstrated by all relevant job positions being in the 'gave advice' or 'gave information' relationships with each task.

Similarly, the degree of differentiation of the system will be a function of the pattern of consultation established. It is to be expected that differentiation will be greater as the number of contributors to each task increases as a function of the complexity of the project. This will, therefore, demand greater intensity of integration.

(iv) *Who exercises what management role as shown by membership of the control loop and who does not, but should do.* The control loops show the degree of continuity between tasks, particularly at decision points, and indicate the level of integration between tasks. The degree of integration exercised over the contributors to each task is identified by the activities of the members of the control loop.

Ideally the control loop composition should show continuity of membership. Normally, the member of the project team responsible for project management and a representative of the client organisation should appear in all control loops and should consistently exercise the same functions. Interruptions of continuity identify potential weaknesses in the integrative functions of the managing system. Interruption of the managing system between tasks in this way occurs particularly when the person in the boundary control relationship does not appear in the same relationship in successive control loops. Such discontinuity may be especially significant at decision points.

Duplication of management functions within a control loop may occur and represents split responsibility between members of the managing system. As well as being undesirable in itself, it also adds to the complexity of the managing system and can impair its effectiveness.

Control loop composition identifies the level of separation of the managing and operating systems. If this separation is not complete, a potential weakness in integration may occur as the objectivity of the person occupying dual roles may be in doubt. If the managing and operating systems are totally undifferentiated, there would be no control loops and all task boxes on the LRA would be occupied by the same job position. If the systems are totally differentiated, none of the task boxes would be occupied by any job position appearing in any control loop.

(v) *The integration of the client's representative with the contributors to the project.* The degree to which the client's representative is involved in the project is demonstrated by his or her presence in the control loops or other relationships. In addition it shows the type of relationship maintained with the tasks, e.g. approval power, general supervision, etc. Consistency of the client's representative is also shown.

Integration of the client's representative and the project team takes place within the tasks and between them. Within a task it is shown when the client's representative appears in the control loop for the task together with the member of the project team responsible for project management. Integration of the client's representative occurs between tasks when the same client's representative appears in the control loops of successive tasks together with the same member of the project team.

Interpretation and use

As can be seen, a particularly interesting aspect of any analysis or design is the composition of the control loops. On projects that use a project manager from outside the client organisation but with close involvement of the client, it is to be expected that all control loops will consist of two members, the project manager and the client's representative. It will be interesting to observe the structure of the control loops if the client's representative has to involve other senior members of the client's organisation at various stages, particularly at key decision points. Continuity may break down and the functions exercised by the duplicated client's representatives in the control loops will generate complex relationships. Although such situations may be unavoidable, the use of LRA to design organisations will enable them to be identified in advance and allow steps to be taken to minimise disruption of the management of the project.

When the members of the project team do not organise themselves positively for managing the project, it is likely that their representative in the control loop will vary, giving lack of continuity between tasks and erratic relationships between the members of the control loops and the contributors who actually undertake the tasks. This is often the case where a senior partner of a professional consultancy is involved in securing a commission at the early stage of a project and then transfers responsibility for managing the project to another member of staff, particularly if that person then continues to maintain an ambiguous role in relation to the management of the project.

The level of detail chosen for the individual tasks and job positions will depend on the purpose for which the LRA is intended. This can range from using it to draw a broadly based map of the whole of a large project down to the design of the way in which a small section of the work is to be organised. Users can pursue the aspect of particular interest to them by selecting and defining the relationships they wish to design or study by the use of appropriately defined matrix symbols. An important advantage is that an LRA can be usefully employed at a level of abstraction suited to construction projects, which are invariably of long duration with a great deal of aggregation of the detailed activities in the project plan and hence in the data available. However, Morris (1994) considers that the technique can be very useful in a variety of project situations. Hughes (1989) has proposed a modification to the LRA which he terms '3R charts'. He believes that in order to handle the data for large complex projects, it is necessary to disaggregate the LRA chart into a series of inter-related components to separate the general from the detail. By doing so the difficulties of complexity in large projects are overcome.

Naturally, any LRA used for design should be capable of being updated and amended in the light of changing environmental conditions. As with any planning tool, it should not remain fixed if conditions change, but should be used to anticipate problems resulting from such changes so that they can be dealt with in the most effective manner.

It should, of course, be pointed out that LRA represents a structural approach to project organisation and that just because the appropriate relationship of contributors, operating system and managing system has been established does not mean that the people involved will work well together in the manner intended, and that the project will be successful. Even if the best organisation has been structured, fundamental aspects of success will be the quality of the skills brought to bear by the contributors, the pattern of authority and power, the leadership of the project team and the attitudes and personal relationships that develop on the project. All of these aspects will be significantly affected by the personal qualities of the managing member of the project team. Nevertheless, the LRA approach should result in an organisation being structured to suit the particular project as the technique demands that the operating system, together with the decision points, are identified before the organisation is fitted to it. That is, the design of the organisation should follow definition of the task to be accomplished. This should give the project manager a sound basis for harnessing the behavioural characteristics of the people involved in the project to the benefit of the client.

Project outcome

Surprisingly, the issue of organisation effectiveness has, in the past, not been of great interest to organisation analysts. They now consider the topic to be important theoretically but find it extremely complex, particularly in terms of the criteria used to judge effectiveness and the manner by which performance against the criteria can be measured (Scott 1998). Understandably, attempting to establish general criteria of effectiveness for the myriad of organistions that exist, is hardly practicable. Many, particularly in the public sector, have diffuse objectives and many claimants, which may generate controversial criteria. Scott identifies three types of indicators of effectiveness: those based on outcomes, on process and on structures.

The major problem of evaluating the effectiveness of any project organisation or any approach to designing organisations is that the success of the organisation in achieving its objective can only be measured against the client's satisfaction with the completed project.

Hence indicators based on outcomes are appropriate. Whilst the issue of identifying the outcome of construction projects may be, on the face of it, more straightforward than for other organisational types, even so there are some significant problems to be faced in establishing acceptable measures of the success of project outcomes. The client's expectation at the very beginning of the project is that it will be fully satisfied with the outcome. The components of the client's satisfaction can be taken as function, including aesthetics, quality, price and time, and the client expects to get the project it wants at the price quoted on the date promised. Whilst such criteria are typically used (Bresnen & Haslam 1991), and are initially appealing they are not without difficulties (Ward *et al.* 1991) such as measuring goal attainment, trade-offs of objectives, goal setting and external factors, particularly if a wide range of stakeholders are to be taken into account.

The project team's objective will be to provide what the client wants and to achieve this they will have to mitigate and harness environmental forces acting upon the project. The object in designing the organisation structure is to provide one which has the greatest capability to produce the project required within the environmental conditions.

Assuming that the people involved in the project have the requisite skills to carry out their work, the success of the project organisation in dealing with the project is the difference between the client's expectation at the beginning of the project and its satisfaction at completion. The greater the uncertainty present on a project, the greater the opportunity for the achieved outcome to fall short of the client's expectation. If the client is completely satisfied with the project on completion, it can be said that the project organisation structure used was suitable for that project in those conditions and performed satisfactorily. The problem is in measuring any shortfall in client satisfaction adequately so that judgements can be made about the performance of organisations that do not produce projects that fully satisfy clients. This is particularly relevant to the value of comparative studies of different organisational approaches. Ideally, objective quantitative means are required for measuring client satisfaction and environmental forces but these are not yet available.

Nevertheless, it is possible to make an assessment, albeit relatively subjectively, of a client's satisfaction with a project outcome and of the strength of environmental influences (Walker & Wilson 1983). This can be useful in making judgements about the success of project organisations. It is important to remember these issues when powerful claims are made in favour of certain organisation arrangements.

For clarity this discussion has made two major assumptions. One is that the client can be readily identified. The other is that it did not relate the evaluation to the time at which it was made.

The client can be difficult to identify for many projects, particularly those in the public sector. In order to obtain a reliable and comprehensive evaluation it is necessary to consult all stakeholders with a claim to make a judgement on a building. For instance, the project team, developer, occupier, owner, workforce, special interest groups and the general public who may use the building, will all have their own viewpoint, each of which is valid from the position which they adopt but all of which contain large elements of subjectivity. The particular objective of any evaluation needs to be clearly identified and the evaluation designed to satisfy it. For example, for clearly identified private clients, building for their own use, the objective would be to evaluate against the brief for the project, even though other stakeholders may well exist. In contrast, for public sector projects it could be that the users believe the brief to have been incorrectly drawn up so the evaluation would need to be against something other than satisfying the brief. The complexity of the issues involved is well illustrated by Liu (1995) who uses an industrial/organisational psychology approach to the evaluation of project outcomes by members of the project team. She identifies the relationship between project success and participant satisfaction in terms of project complexity and commitment using the hypothesis 'the effects of project success on participant satisfaction [are] dependent on project complexity and the valence of project success is dependent on the level of aspiration of participant satisfaction as well as project complexity'.

An aspect which causes problems with evaluations is that as time passes, people's perceptions of a building change. In the early phases of planning and feasibility, a building exists only as an idea and reactions to it are to an abstract concept which is only defined in very approximate terms. As the concept of the building is developed through drawings and specifications of materials, reactions are likely to be more detailed but still only related to a model of the building, not reality. It is also possible that reactions to the building could be influenced by the construction process itself in terms of the problems and inconveniences it may have generated.

Reactions to a building immediately upon occupation tend to be exaggerated. Expectations are high, the building is in pristine condition, users are unfamiliar with how the building should be used and there are frequently teething troubles with technical aspects. The building then matures, it becomes taken for granted by users, owners, and the general public, it is altered and adapted as demands on it change, it may be rehabilitated but it eventually moves towards the end of its economic life. At each of these stages perceptions of the building will be judged against different criteria by the different classes of people with an interest in it.

This scenario is recounted to impress that, in evaluating the outcome of projects, it is necessary to spell out clearly the perspective being taken (e.g. user, owner, project team, special interest group) and the criteria against which the evaluation is being carried out. It also illustrates the problems and perhaps, in-appropriateness of attempting to arrive at a single measure of the outcome of a project across all perspectives of a building.

Criticism of function, cost and time as performance measures for project man-agers is often heard. For example, Dainty *et al.* (2003) believe that these measures are too crude. They believe that traditional success parameters should be rede-fined 'to consider the knowledge, skills and behavioural inputs which contribute to superior performance. These can then be used to engender a more appro-priate set of management behaviours that lead to improved project outcomes'. Through discussions with focus groups and statistical analysis, they identify 'sets of criteria for management performance (success factors) that must be displayed by contractors' project managers in order that they achieve typical project ob-jectives (project success)'. These criteria were found to be: team building; lead-ership; decision making; mutuality and approachability; honesty and integrity; communication; learning, understanding and application; self-efficacy; and ex-ternal relations. Whilst these factors may well contribute to performance, the study did not attempt to correlate them to the achievement of the conventional project success criteria. Although project managers' performance can be mea-sured against behavioural criteria that may be valuable for purposes internal to their employing organisation, clients and other stakeholders are nevertheless more likely to be interested in how well the project meets the conventional success criteria.

Presentation of project organisations

Increasingly, throughout the world, invitations to bid for contracts require bid-ders to submit details of their proposed organisation structure and approach to the project in addition to the bid price. Quite reasonably, clients wish to be as-sured that the bidder has a sound and realistic approach to the organisation and successful completion of the project. The problem faced by the bidder, assuming that an acceptable strategy for the project can be produced, is that of convincing the client that the bidder can organise successfully to achieve what is claimed. Propositions that adopt a traditional approach will not necessarily be accepted on face value as many overseas clients will not be familiar with the style and terminology. Frequently, such an approach will not be compatible with the client's needs or the local environment. In many cases, a technique is needed that will explain clearly and convincingly to clients how the project is to be tackled.

Obviously, an LRA will not by itself fully satisfy the client's requirements but it can provide a powerful illustration to the client of how the bidder proposes to organise to carry out the project. It also provides a valuable focal point for the co-ordination of other documentation which will be required to convince the client that the bidder has the capability to complete the work successfully. For example, it is usual for the client to require details of the firms and personnel the bidder intends to involve in the project. Such descriptions can be linked to the LRA by cross-referencing their descriptions to their positions on the LRA. The LRA also follows the style of a network and if it is intended to submit a network with the bid, this can also be co-ordinated and cross-referenced to the LRA. Such a package can look very convincing to a client and the client's confidence in the bid can increase, with a consequently greater chance of success.

The large variety of organisation approaches and contractual arrangements which are available and necessary to solve the wide range of different projects on offer throughout the world make it essential for bidders to adopt a creative and constructive method of putting their approaches to clients. The LRA can be used to illustrate any of these approaches because of its flexibility in being able to represent organisations at various levels of detail. It is therefore suitable for bids for construction only, for design-and-build bids and for design-only bids. It is particularly valuable for joint ventures as it will clearly show the aspects of the projects in which the various partners are to be involved. The definitions of the matrix symbols will be most important in such arrangements, as they will define the responsibilities and authority of the various participants.

A particular advantage of the LRA technique in the presentation of proposals is that it should explain to clients what their level of involvement in the project is expected to be. This can enable detailed discussions to be held with the client prior to the bid being submitted, which will be advantageous to the bidder in clarifying the client's requirements in terms of both the project brief and of the client's expectation of involvement in the decision-making process and organisation of the project. If, upon receiving a bid, a client can clearly see not only the design and the price but also how they are to be achieved and what is expected of the client in the process, the client is likely to be assured that the bidder is a professional and competent organisation.

References

Abidin, N. & Pasquire, C. (2005) Delivering sustainability through value management: concept and performance overview. *Engineering, Construction and Architectural Management,* **12**, 2.

Ackoff, R.L. (1969) Systems, organisation and interdisciplinary research. In: Emery, F.E. (ed.) *Systems Thinking.* London: Penguin.

Ackoff, R.L. (1971) Towards a system of system concepts, *Management Science,* **17**, 661–71.

Agger, B. (1991) Critical theory, poststructuralism, postmodernism: their sociological relevance. *Annual Review of Sociology,* **17**, 105–31.

Akintoye, A. (1994) Design and build: a survey of construction contractors' views. *Construction Management and Economics,* **12**, 155–63.

Akintoye, A. & Fitzgerald, E. (1995) Design and build: a survey of architects' views. *Engineering, Construction and Architectural Management,* **2**, 1.

Akintoye, A., Hardcastle, C., Beck, M., Chinyio, E. & Asenova, D. (2003) Achieving best value in private finance initiative project procurement. *Construction Management and Economics,* **21**, 461–70.

Aldrich, H. (2000) *Organisations Evolving.* London: Sage.

Allen, D. (1984) Towards the client's objective. In: Brandon, P.S. & Powell, J.A. (eds) *Quality and Profit in Building Design.* London: Spons.

Allen, P.M. (2001) A complex systems approach to learning in adaptive networks. *International Journal of Innovation Management,* **5**, 2.

Alvesson, M. & Deetz, S. (1996) Critical theory and postmoderism approaches to organisational studies. In: Clegg, S., Hardy, C. & Nord, W. (eds.) *Handbook of Organisation Studies.* Thousand Oaks, CA: Sage.

Ancona, D., Kockan, T., Scully, M., Van Maanen, J. & Westney, D. E. (1996) *Managing for the Future: Organizational Behavior and Processes.* Cincinnati, OH: Southwestern College Publishing.

Anderson, S.D. (1992) Project quality and project managers. *International Journal of Project Management,* **10**, 138–44.

Ankrah, N. & Langford, D. (2005) Architects and contractors: a comparative study of organisational cultures. *Construction Management and Economics,* **23**, 595–607.

Anon (1955) How to know who does what. *Mill and Factory,* January, 75–8.

Association of Project Managers (1984) *Closing the Gaps in Project Management Systems.* London: Butterworths.

Baiden, B., Price, A. & Dainty, A. (2006) The extent of team integration within construction projects. *International Journal of Project Management,* **24**, 1.

Barley, S. & Kunda, G. (2001) Bringing work back in. *Organizational Science,* **12**, 76–95.

Barnard, I. (1938) *The Functions of the Executive.* Cambridge, MA: Harvard University Press.

Barney, J.B. (1990) The debate between traditional management theory and organisational economics: substantive differences or intergroup conflicts? *Academy of Management Review*, **15**, 382–93.

Barrett, P.S., Hudson, J. & Stanley, C. (1996) Is briefing innovation? In: *The Organization and Management of Construction: Shaping Theory and Practice*, Langford, A. & Retik, A. (eds) Vol. 3. London: E&FN Spon.

Barrett, P. & Stanley, C. (1999) *Better Construction Briefing*. Oxford: Blackwell Science.

Bennett, J. (1985) *Construction Project Management*. Cambridge: Butterworth Publishing.

Bennett, J. (1991) *International Construction Project Management: General Theory and Practice*. Oxford: Butterworth-Heinemann.

Bennigson, L.A. & Balthasas, H.V. (1974) Forecasting coordination problems in pharmaceutical research and development. In: *1974 Proceedings of the Project Management Institute*. Paris: Internet.

Bennis, W.G. (1959) Leadership theory and administrative behaviour. *Administrative Science Quarterly*, **4**, 259–301.

Bertalanffy, L. von (1969) *General Systems Theory: Essays on its Foundation and Development*. New York: Braziller.

Betts, M. & Wood-Harper, T. (1994) Re-engineering construction: a new management research agenda. *Construction Management and Economics*, **12**, 551–6.

Blake, R.R. & Mouton, J.S. (1978) *The New Managerial Grid*. Houston: Gulf Publishing.

Blau, P. (1970) A formal theory of differentiation in organisations. *American Sociological Review*, **35**, 2.

Blismas, N., Sher, W., Thorpe, A. & Baldwin, A. (2004) Factors influencing project delivery within construction clients' multi-project environments. *Engineering, Construction and Architectural Management*, **11**, 2.

Blockley, D. & Godfrey, P. (2000) *Doing it Differently: Systems for Rethinking Construction*. London: Thomas Telford.

Bonoma, T.V. & Slevin, D. (1978) *Executive Survival Manual*. Belmont: Wadsworth Publishing Co.

Boulding, K.E. (1956) General systems theory: the skeleton of science. *Management Science*, **2**, 197–208.

Bowen, P.A., Hindle, R.D. & Pearl, R.G. (1997) The effectiveness of building procurement systems in the attainment of client objectives. In *Proceedings of W-92 Conference on Procurement Systems*, C.I.B. Publication No. 203.

Bowley, M. (1966) *The British Building Industry*. Cambridge: Cambridge University Press.

Bresnen, M.J. (1991) Construction contracting in theory and practice: a case study. *Construction Management and Economics*, **9**, 247–63.

Bresnen, M.J. & Haslam, C.D. (1991) Construction industry clients: a survey of their attributes and project management practices. *Construction Management and Economics*, **9**, 327–42.

Bresnen, M. & Marshall, N. (2000a) Partnering in construction: a critical review of issues, problems and dilemmas. *Construction Management and Economics*, **18**, 229–37.

Bresnen, M. & Marshall, N. (2000b) Motivation, commitment and the use of incentives in partnerships and alliances. *Construction Management and Economics*, **18**, 587–98.

Bresnen, M. & Marshall, N. (2000c) Building partnerships: case studies of client–contractor collaboration in the UK construction industry. *Construction Management and Economics*, **18**, 819–32.

Bridge, A. & Tisdell, C. (2004) The determinants of the vertical boundaries of the construction firm. *Construction Management and Economics*, **22**, 807–25.

Bridge, A. & Tisdell, C. (2006) The determinants of the vertical boundaries of the construction firm: a response. *Construction Management and Economics*, **24**, 233–36.

Briscoe, G., Dainty, A., Millett, S. & Neale, R. (2004) Client-led strategies for construction supply chain improvement. *Construction Management and Economics*. **22**, 193–201.

British Property Federation (1983) *Manual of the BPF System*. London: British Property Federation.

British Standards Institution (2000) *Guide to Project Management*. *BS6079*. London: BSI.

Brown, A. (1998) *Organisational Culture*. London: Financial Times Management.

Buchko, A. (1994) Conceptualisation and measurement of environmental uncertainty: an assessment of the Miles and Snow Perceived Environmental Uncertainty Scale. *Academy of Management Journal*, **37**, 410–25.

Buckley, W. (1968) 'Society as a complex adaptive system: In: Buckley, W. (ed.) *Modern Systems Research for the Behavioural Scientist: A Sourcebook*. Chicago: Aldine.

Burke, W.W. (1986) Leadership as empowering others. In: Srivastna, S. (ed.) *Executive Power*. San Francisco: Jossey-Bass.

Burns, R. & Stalker, G.M. (1966) *The Management of Innovation*. London: Tavistock Publications.

Cameron, K. & Quinn, R. (1999) *Diagnosing and Changing Organisational Culture*. Reading, MA: Addison-Wesley Longman.

Campbell, J.P., Dunnette, M.D., Lawler, E.E., Weick, K.E., (1970) *Managerial Behaviour, Performance and Effectiveness*. New York: McGraw-Hill.

Carpenter, J.B.G. (1981) The UK system of construction procurement and what is wrong: how to improve? Paper to *RICS Quantity Surveyors' 12th Triennial Conference*. London: RICS.

Carrizosa, A. & Ortegon, M. (1998) Using systems metaphors to interpret the edge of chaos. In: *Systems Science and Engineering – Third International Conference, Beijing*.

Chang, C-Y. (2006a) The principle of inconsistent trinity in the selection of procurement systems. London: University College London.

Chang, C-Y. (2006b) The determinants of the vertical boundaries of the construction firm: Comment. *Construction Management and Economics*, **24**, 222–39.

Chapman, R.J. (1999) The likelihood and impact of changes of key project personnel on the design process. *Construction Management and Economics*, **17**, 99–106.

Chartered Institute of Building (1979) *Project Management in Building*. Occasional Paper No. 20. Ascot: The Chartered Institute of Building.

Chartered Institute of Building (2002)*Code of Practice for Project Management for Construction and Development*. Oxford: Blackwell.

Chau, K.W. & Walker, A. (1994) Institutional costs and the nature of subcontracting in the construction industry. In: Rowlinson, S.M. (ed.) *Proceedings of CIB W92 Symposium: East Meets West: Procurement Systems*. CIB Publication No. 175. Hong Kong: Department of Surveying, University of Hong Kong.

Checkland, P. (1989) Soft systems methodology. In: *Rational Analysis for a Problematic World*, Rosenhead, J. (ed.). Chichester: Wiley.

Cherns, A.B. & Bryant, D.T. (1984) Studying the client's role in construction management. *Construction Management and Economics*, **2**, 177–84.

Cheung, S.N.S. (1983) The contractual nature of the firm. *Journal of Law and Economics*, **26**, 1–21.

Cheung, S.N.S. (1992) On the New Institutional Economics. In: *Contract Economics*, Werin, L. & Wijkander, H. (eds). Oxford: Blackwell.

Child, J. (1972) Organisational structure, environment and performance. The role of strategic choice. *Sociology*, **6**, 1–22.

Child, J. (1977) *Organisation*. London: Harper & Row.

Chiles, T.H. & McMackin, J.F. (1996) Integrating variable risk preferences, trust, and transaction cost economics. *Academy of Management Review*, **21**, 1.

Chinyio, E.A., Olomolaiye, P.O., Kometa, S.T. & Harris, F.C. (1998) A needs-based methodology for classifying construction clients and selecting contractors. *Construction Management and Economics*, **16**, 91–8.

Cicmil, S. & Marshall, D. (2005) Insights into collaboration at the project level: complexity, social interactions and procurement mechanisms. *Building Research and Information*, 33, 6.

Cleland, D.I. & King, W.R. (1972) *Management: A Systems Approach*. New York: McGraw-Hill.

Cleland, D.I. & King, W.R. (1983) *Systems Analysis and Project Management*. New York: McGraw-Hill.

Coase, R.H. (1937) The nature of the firm. *Economica*, n.s., 4, 386–405.

Coase, R.H. (1988) *The Firm, the Market and the Law*. Chicago: The University of Chicago Press.

Coase, R.H. (1991) *The Institutional Structure of Production*. Stockholm: The Royal Swedish Academy of Sciences.

Coase, R.H. (1991) The nature of the firm: influence. In: Williamson. O. & Winter, S. (eds) *The Nature of the Firm: Origins, Evolution, and Development*. New York: Oxford University Press.

Cohen, S.G. & Bailey, D.E. (1997) What makes teams work: Group effectiveness research from the shop floor to the executive suite. *Journal of Management*, 23, 3.

Conger, J.A. & Kanungo, R.N. (1988) The empowerment process: integrating theory and practice. *Academy of Management Review*, 13, 471–82.

Constantino, N., Pietroforte, R. & Hamill, P. (2001) Subcontracting in commercial and residential construction: an empirical investigation. *Construction Management and Economics*, 19, 439–47.

Construction Industry Council (2001) *Construction Project Management Skills*. London: Construction Industry Council.

Construction Industry Institute (1989) *Partnering: Meeting the Challenges of the Future*. Texas: Construction Industry Institute (CII).

Cox, A. & Ireland, P. (2002) Managing construction supply chains: the common sense approach. *Engineering, Construction and Architectural Management*, 9, 5/6.

Cox, A. & Townsend, M. (1997) Latham as half-way house: a relational competence approach to better practice in construction procurement. *Engineering, Construction and Architectural Management*, 4, 2.

Culp, G. (2004) *The Lead Dog Has the Best View: Leading Your Project Team to Project Success*. Reston, VA: ASCE.

Dahlman, C.J. (1979) The problem of externality. *The Journal of Law and Economics*, 22, 141–62.

Dainty, A., Briscoe, G. & Millett, S. (2001) Subcontractor perspectives on supply chain alliances. *Construction Management and Economics*, 19, 841–8.

Dainty, A., Bryman, A., Price, A., Geasley, K., Soetanto, R. & King, N. (2005) Project affinity: the role of emotional attachment in construction projects. *Construction Management and Economics*, 23, 241–4.

Dainty, A., Cheng, M. & Moore, D. (2003) Redefining performance measures for construction project mangers: an empirical evaluation. *Construction Management and Economics*, 21, 209–18.

Dalton, G.W., Lawrence, P.C. & Lorsch, J.W. (1970) *Organisational Structure and Design*. New York: Irwin.

Daniel, D.W. (1990) Hard problems in a soft world. *International Journal of Project Management*, 8, 79–83.

Das, T.K. & Teng, B. (1998) Between trust and control: Developing confidence in partner cooperation in alliances. *Academy of Management Review*, 23, 3.

Demsetz, H. (1991) The theory of the firm revisited. In: Williamson, O.E. & Winter, S.G. (eds) *The Nature of the Firm*. Oxford: Oxford University Press.

Department of Industry (1982) *The United Kingdom Construction Industry*. London: HMSO.

Department of the Environment (1995) *Housing and Construction Statistics*. London: HMSO.

DiMaggio, P. & Powell, W. (1983) The iron cage revisited: institutional isomorphism and collective rationality in organisation fields. *American Sociological Review*, **48**, 147–60.

Dirks, K. & Skarlicki, D. (2004) Trust in leaders: Existing research and emerging issues. In: Kramer, R. & Cook, K. (eds) *Trust and Distrust in Organisations*. New York: Russell Sage Foundation.

Donaldson, L. (1990a) The etheral hand: organisational economics and management theory. *Academy of Management Review*, **15**, 369–81.

Donaldson, L. (1990b) A rational thesis for criticisms of organisational economics: a reply to Barney. *Academy of Management Review*, **15**, 394–401.

Donaldson, L. (1996) *For Positivist Organisation Theory*. Thousand Oaks, CA: Sage.

Dornbusch, S.M. & Scott, W.R. (1975) *Evaluation and the Exercise of Authority*. San Francisco: Jossey-Bass.

Egan, J. (1998) *Rethinking Construction*. London: The Department of the Environment, Transport and the Regions.

Eilon, S. (1979) *Aspects of Management*. Oxford: Pergamon Press.

Emerson, R.M. (1962) Power-dependence relations. *American Sociological Review*, **27**, 31–40.

Emery, F.E. (1959) *Characteristics of Socio-Technological Systems*. Tavistock Document 527. London: Tavistock Publications.

Emery, F.E. & Trist, E.L. (1965) The causal texture of organisational environment. *Human Relations*, **18**, 21–32.

Emirbayer, M. (1997) Manifesto for a relational sociology. *American Journal of Sociology*, **103**, 2.

Eskrod, P. & Blichfeldt, B. (2005) Managing team entries and withdrawals during the project life cycle. *International Journal of Project Management*, **23**, 7.

Etzioni, A. (1964). *Modern Organisations*. Englewood Cliffs, NJ: Prentice Hall.

Fayol, H. (1949 trans.) *General and Industrial Management*. London: Pitman (First published in 1919).

Fellows, R. (2006) Culture. In: Lowe, D. & Leiringer, R. (eds) *Commercial Management of Projects: Defining the Discipline*. Oxford: Blackwell.

Fergusson, H. & Langford, D.A. (2006) Strategies for managing environmental issues in construction organisations. *Engineering, Construction and Architectural Management*, **13**, 2.

Fernie, S., Green, S. & Weller, S. (2003) Dilettantes, discipline and discourse: requirements of management for construction. *Engineering, Construction and Architectural Management*, **10**(5).

Fewings, P. (2005) *Construction Project Management: An Integrated Approach*. Abingdon: Taylor & Francis.

Fielder, F.E. (1967) *A Theory of Leadership Effectiveness*. New York: McGraw-Hill.

Fincham, R. & Rhodes, P. (2005) *Principles of Organizational Behaviour*. New York: Oxford University Press.

Finkelstein, S. (1992) Power in top management teams: dimensions, measurement and validation. *Academy of Management Journal*, **35**, 505–38.

Flood, R.L. (1999) *Rethinking 'The Fifth Discipline': Learning Within the Unknowable*. London: Routledge.

French, W. & Bell, C. (1990) *Organisation Development*. Englewood Cliffs, NJ: Prentice-Hall.

Fryer, B. (1985) *The Practice of Construction Management*. Oxford: Blackwell Science.

Fryer, B., Egbu, C., Ellis, R. & Gorse, C. (2004) *The Practice of Construction Management*. Oxford: Blackwell.

Galbraith, J.K. (1967) *The New Industrial State*. Boston: Houghton Mifflin.

Gilbert, G.P. (1983) Styles of project management. *International Journal of Project Management*. **1**, 189–93.

Gleick, J. (1993) *Chaos*. London: Abacus.

Goodacre, P., Pain, J., Murray, J. & Noble, M. (1982) *Research in Building Design*. Occasional Paper No. 7. Reading: Department of Construction Management, University of Reading.

Gray, C. (2004) Technology clusters used to achieve project performance change. In: Pantouvakis, J. (ed.) *Proceedings of the Third Scientific Conference on Project Management: Clustering in Construction Project Management*. Athens: National Technical University of Athens.

Green, S.D. (1994) Sociological paradigms and building procurement. In: Rowlinson, S.M. (ed.) *Proceedings of CIB W92 Symposium: East Meets West: Procurement Systems*. CIB Publications No. 175. Hong Kong: Department of Surveying, University of Hong Kong.

Green, S.D. (1996) A metaphorical analysis of client organizations and the briefing process. *Construction Management and Economics*, **14**, 155–64.

Green, S.D. (1998) The technocratic totalitarianism of construction process improvement: a critical perspective. *Engineering, Construction and Architectural Management*, **5**, 4.

Green, S.D. (1999) Partnering: The propaganda of corporatism. *Journal of Construction Procurement*, **5**, 2.

Green, S. D. & May, S. (2005) Lean construction: arenas of enactment, models of diffusion and the meaning of 'leanness'. *Building Research & Information*, **33**, 6.

Green, S.D. & Simister, S.J. (1999) Modeling client business processes as an aid to strategic briefing. *Construction Management and Economics*, **17**, 63–76.

Greening, D.W. & Gray, B.L. (1994) Testing a model of organisational response to social and political issues. *Academy of Management Journal*, **37**, 467–98.

Gresov, C. & Drazin, R. (1997) Equifinality: Functional equivalence in organization design. *Academy of Management Review*, **22**, 2.

Griesinger, D.W. (1990) The human side of economic organisation. *Academy of Management Review*, **15**, 478–99.

Gulick, L. & Urwick, L. (eds) (1937) *Papers on the Science of Administration*. New York: Institute of Public Administration, Columbia University.

Hammer, M. & Champy, J. (1993) *Re-engineering the Corporation: A Manifesto for Business Revolution*. London: Harper Collins.

Hammuda, I. & Dulaima, M.F. (1997) The theory and application of empowerment in construction: A comparative study of the different approaches to empowerment in construction, service and manufacturing industries. *International Journal of Project Management*, **15**, 5.

Handler, A.B. (1970) *Systems Approach to Architecture*. Amsterdam: Elsevier.

Handy, C. (1985) *Understanding Organisations*. Harmondsworth: Penguin.

Hardcastle, C. & Boothroyd, K. (2003) Risks overview in public private partnerships. In: Akintoye, A., Beck, M. & Hardcastle, C. (eds) *Public Private Partnerships*. Oxford: Blackwell.

Harvey-Jones, J. (1993) *Managing to Survive*. London: Heinemann.

Hennart, J.F. (1994) The comparative institutional theory of the firm: some implications for corporate strategy. *Journal of Management Studies*, **31**, 193–207.

Hersey, P. & Blanchard, K.H. (1972) *Management of Organisation Behaviour*. New Jersey: Prentice-Hall.

Herzberg, F. (1968) *Work and the Nature of Man*. London: Staples Press.

Hesterly, W.S., Liebestkind, J. & Zenger, T.R. (1990) Organisational economics: an impending revolution in organisation theory? *Academy of Management Review*, **15**, 402–420.

Hickson, D.J., Hinings, C.R., Lee, C.A., Schreck, R.E. & Pennings, J.M. (1971) A strategic contingencies' theory of intraorganisational power. *Administrative Science Quarterly*, **16**, 216–29.

Higgins, G. & Jessop, N. (1965) *Communication in the Building Industry*. London: Tavistock Publications.

Hillebrand, P. (2000) *Economic Theory and the Construction Industry.* Hampshire: Macmillan Press.

Hinings, B. (2003) Organisation structure. In: Westwood, R. & Clegg, S. (eds) *Debating Organisations: Point-Counterpoint in Organisation Studies.* Oxford: Blackwell.

HMSO (1944) *Report of the Management and Planning of Contracts (The Simon Report).* London: HMSO.

HMSO (1950) *Report of the Working Party on the Building Industry (The Phillips Report).* London: HMSO.

HMSO (1962) *Survey of the Problems Before the Construction Industries (The Emmerson Report).* London: HMSO.

HMSO (1964) *The Placing and Management of Contracts for Building and Civil Engineering Works (The Banwell Report).* London: HMSO.

HMSO (1967) *A Survey of the Implementation of the Recommendations of the Committee under the Chairmanship of Sir Harold Banwell on the Placing and Management of Contracts for Building and Civil Engineering Works (Action on the Banwell Report).* London: HMSO.

Hofstede, G. (1994) The business of international business is culture.*International Business Review*, 3, 1.

Hopkins, W., Hopkins, S. & Mallette, P. (2005)*Aligning Organisational Subcultures for Competitive Advantage: A Strategic Change Approach.* New York: Basic Books.

Howell, J. & Costly, L. (2006) *Understanding Behaviours for Effective Leadership.* Upper Saddle River, NJ: Pearson Education.

Hughes, W. (1997) Construction management contracts: law and practice. *Engineering, Construction and Architectural Management*, 4, 1.

Hughes, W. (2003) A comparison of two editions of the RIBA Plan of Work. *Engineering, Construction and Architectural Management*, 10, 5.

Hughes, W.P. (1989) *Organisational Analysis of Building Projects.* PhD Thesis, Liverpool: Liverpool Polytechnic (now Liverpool John Moores University).

Ireland, V. (1985) The role of managerial actions in the cost, time and quality performance of high rise commercial building projects. *Construction Management and Economics*, 3, 59–87.

Jackson, M.C. (2000) *Systems Approaches to Management.* London: Kluwer Academic.

Janis, I.L. (1983) *Groupthink.* Boston: Houghton Mifflin.

Jawahar-Nesan, L. & Price, A.D.F. (1997) Formulation of best practices for owner's representatives. *Journal of Management in Engineering*, 13, 1.

Jensen, M.C. & Meckling, W.Y. (1976) Theory of the firm: managerial behaviour, agency costs, and ownership structure. *Journal of Financial Economics*, 3, 305–60.

Johnson, J.L. & Burton, B.K. (1994) Chaos and complexity theory for management: caveat emptor. *Journal of Management Enquiry*, 3, 320.

Johnson-George, C. & Swap, W. (1982) Measurement of specific personal trust: Construction and validation of a scale to assess trust in a specific other. *Journal of Personality and Social Psychology*, 43, 1306–17.

Kadfors, A. (2004) Trust in project relationships – inside the black box. *International Journal of Project Management*, 22, 3.

Kakabadse, A., Bank, J. & Vinnicombe, S. (2004) *Working in Organisations.* Aldershot: Gower.

Kanter, R.M. (1977) *Men and Women of the Corporation.* New York: Basic Books.

Kanter, R.M. (1983) *The Change Masters.* London: Unwin.

Kast, F.E. & Rosenzweig, J.E. (1985) *Organisation and Management: A Systems and Contingency Approach.* New York: McGraw-Hill.

Katz, D. & Kahn, R.L. (1978) *The Social Psychology of Organisations.* New York: John Wiley & Sons.

Keegan, A. & Den Hartog, D. (2004) Transformational leadership in a project-based environment: a comparative study of the leadership styles of project managers and line managers. *International Journal of Project Management*, 22, 8.

Kilbert, C., Sendzimir, J. & Guy, B. (2000) Construction ecology and metabolism: natural system analogues for a sustainable built environment. *Construction Management and Economics*, **18**, 903–16.

Kirkman, B.L. & Rosen, B. (1999) Beyond self-management: antecedents and consequences of team empowerment. *Academy of Management Journal*, **42**, 1.

Knowles, R. (1986) Project manager: the legal position. *The Chartered Quantity Surveyor.* July (Part 1), 11; August (Part 2), 10–11.

Knox, F. & Hennesey, J. (1966) *Restrictive Practices in the Building Industry.* London: Institute of Economic Affairs.

Kometa, S.T., Olomolariye, P.O. & Harris, F.C. (1994) Attributes of UK construction clients influencing project consultants' performance. *Construction Management and Economics*, **12**, 433–43.

Koontz, H. (1961) The management theory jungle. *Journal of the Academy of Management*, December.

Koskela, L. & Ballard, G. (2006) Should project management be based on theories of economics or production? *Building Research and Information*, **34**, 2.

Lai, L. (2000) The Coasian market-firm dichotomy and subcontracting in the construction industry. *Construction Management and Economics*, **18**, 355–62.

Lane, D. & Jackson, M.C. (1995) Only connect! An annotated bibliography reflecting the breadth and diversity of systems thinking. *Systems Research*, **12**, 217.

Lansley, P. (1994) Analysing construction organisations. *Construction Management and Economics*, **12**, 337–48.

Larson, C.E. & LaFasto, F.M.J. (1989) *Teamwork: What must go right/What can go wrong.* Newbury Park, CA: Sage.

Larson, E. (1995) Project partnering: results of study of 280 construction projects. *Journal of Management in Engineering*, **11**, 30–35.

Latham, M. (1994) *Constructing the Team.* London: HMSO.

Lawrence, P.C. & Lorsch, J.W. (1967) *Organisation and Environment: Managing Differentiation and Integration.* Boston: Graduate School of Business Administration, Harvard University.

Lazar, F.D. (2000) Project partnering: Improving the likelihood of win/win outcomes. *Journal of Management in Engineering*, **16**, 2.

Leavitt, H. (2005)*Top Down: Why Hierarchies are Here to Stay and How to Manage Them More Effectively.* Boston, MA: Harvard Business School Press.

Leigh, A. (1983) *Decisions, Decisions! A Practical Guide to Problem Solving and Decision Making.* Aldershot: Gover.

Levy, S. (2000) *Management in Construction.* New York: McGraw-Hill.

Li, B. & Akintoye, A. (2003) An overview of public private partnerships. In: Akintoye, A., Beck, M. & Hardcastle, C. (eds) *Public Private Partnerships.* Oxford: Blackwell.

Li, B., Akintoye, A., Edwards, P.J., & Hardcastle, C. (2005) Critical success factors for PPP/PFI projects in the UK construction industry. *Construction Management and Economics*, **23**, 459–71.

Li, H., Cheng, E.W.L. & Love, P.E.D. (2000) Partnering research in construction. *Engineering Construction and Architectural Management*, **7**, 1.

Likert, R. (1961) *New Patterns of Management.* New York: McGraw-Hill.

Lim, E. & Ling, F. (2002) Model for predicting clients' contribution to project success. *Engineering, Construction and Architectural Management*, **9**, 5/6.

Lingard, H., Hughes, W. & Chinyio, E. (1998) The impact of contractor selection method on transaction costs: a review. *Journal of Construction Procurement*, **4**, 2.

Liu, A. & Fang, Z. (2006) The power paradigm of project leadership. *Construction Management and Economics*, **24**, 497–507.

Liu, A., Fellows, R. & Fang, Z. (2003) The power paradigm of project leadership. *Construction Management and Economics*, **21**, 819–29.

Liu, M.M.A. (1995) *Evaluation of the Outcome of Construction Projects.* PhD Thesis, Hong Kong: University of Hong Kong.

Liu, M.M.A. (1999) A research model of project complexity and goal commitment effects on project outcome. *Engineering, Construction and Architectural Management*, **6**, 2.

Liu, M.M.A. & Fellows, R. (2001) An Eastern perspective on partnering. *Engineering, Construction and Architectural Management*, **8**, 1.

London, K. & Kenley, R. (2001) An industrial organization economic supply chain approach for the construction industry: a review. *Construction Management and Economics*, **19**, 777–88.

Loosemore, M. (1994) Problem behaviour. *Construction Management and Economics*, **12**, 511–20.

Loosemore, M. (1999) Responsibility, power and construction conflict. *Construction Management and Economics*, **17**, 699–709.

Loosemore, M. & Tan, C.C. (2000) Occupational stereotypes in the construction industry. *Construction Management and Economics*, **18**, 559–566.

Lorsch, J.W. (1979) Making behavioural science more useful. *Harvard Business Review*, **57**, 171.

Love, P.E.D., Skitmore, M. & Earl, G. (1998) Selecting a suitable procurement method for a building project. *Construction Management and Economics*, **16**, 221–33.

Love, P., Irani, Z., Cheng, E. & Li, H. (2002) A model for supporting inter-organizational relations in the supply chain. *Engineering, Construction and Architectural Management*, **9**, 1.

Lovell, R.J. (1993) Power and the project manager. *International Journal of Project Management*, **11**, 73–8.

Lutzkendorf, T. & Lorenz, D. (2005) Sustainable property investment: valuing sustainable buildings through property performance assessment. *Building Research and Information*, **33**, 3.

Lycett, M., Rassau, A. & Danson, J. (2004) Programme management: a critical review. *Engineering, Construction and Architectural Management* **22**, 4.

Maloney, W.F. & Federle, M.O. (1991) Organisation, culture and management. *Journal of Management in Engineering*, **7**, 43–58.

Manager Update (1997) **9**, 2.

Manz, C.C. & Neck, C.P. (1995) Teamthink: beyond the groupthink syndrome in self-managing work teams. *Journal of Managerial Psychology*, **10**, 7–15.

March, T.G. & Simon, H.A. (1958) *Organisation*. New York: John Wiley & Sons.

Maslow, A.H. (1954) *Motivation and Personality*. New York: Harper.

Masterman, J.W.E. & Gameson, R.N. (1994) Client characteristics and needs in relation to their selection of building procurement systems. In: Rowlinson, S.M. (ed.) *Proceedings of CIB92 Symposium: East Meets West: Procurement Systems*. CIB Publication No. 175. Hong Kong: Department of Surveying, University of Hong Kong.

Matthews, J., Tyler, A. & Thorpe, A. (1996) Pre-construction project partnering. *Engineering Construction and Architectural Management*, **3**, 1/2.

Mayer, R.C., Davis, J.H. & Schoorman, F.D. (1995) An integrative model of organizational trust. *Academy of Management Review*, **20**, 3.

McCabe, S., Rooke, J., Seymore, D. & Brown, P. (1998) Quality managers, authority and leadership. *Construction Management and Economics*, **16**, 447–57.

McDermott, P., Khalfan, M. & Swan, W. (2005) 'Trust' in construction projects. In: Lowe, D. & Emsley, M. (eds) *International Commercial Management Symposium*. Manchester: University of Manchester.

McGregor, D. (1960) *The Human Side of Enterprise*. New York: McGraw-Hill.

McKinley, W., Mone, M.A. & Moon, G. (1999) Determinants and development of schools in organization theory. *Academy of Management Review*, **24**, 4.

McMillan, E. (2004) *Complexity, Organisations and Change*. London: Routledge.

Meindl, J. (1990) On leadership: An alternative to the conventional wisdom. In: Cummings, M. & Staw, B. (eds), *Research in Organisational Behaviour*, Vol. 12. Greenwich, CT: JAI Press.

Meyer, J.W. (1994) Rationalised Environments. In: Scott, R.W. & Meyer, J.W. (eds) *Institutional Environments and Organisations: Structural Complexity and Individualism*. Thousand Oaks, CA: Sage.

Meyer, J.W. & Rowan, B. (1977) Institutionalised organisations: formal structure as myth and ceremony. *American Journal of Sociology*, **83**, 340–63.

Meyer, J.W. & Scott, W.R. (1983). *Organisational Environments: Ritual and Rationality*. Beverly Hills, CA: Sage.

Miles, R.E. & Snow, C.C. (1978) *Organisational Strategy, Structure, and Process*. New York: McGraw-Hill.

Miller, C., Packham, G. & Thomas, B. (2002) Harmonization between main contractors and subcontractors: a prerequisite for lean construction. *Journal of Construction Research*, **3**, 1.

Miller, E.J. (1959) Technology, territory and time: the internal differentiation of complex production systems. *Human Relations*, **12**, 243–72.

Miller, E.J. & Rice, A.K. (1967) *Systems of Organisation: the Control of Task and Sentient Boundaries*. London: Tavistock Publications.

Mintzberg, H. (1973) *The Nature of Managerial Work*. New York: Harper and Row.

Mintzberg, H. (1979) *The Structure of Organisations*. Englewood Cliffs, NJ: Prentice-Hall.

Mintzberg, H. (1989) *Mintzberg on Management: Inside Our Strange World of Organisations*. New York: Free Press.

Mobbs, G.N. (1976) *Industrial Investment – A Case Study in Factory Building*. Slough: Slough Estates Ltd.

Mohsini, R.A. & Davidson, C.H. (1992) Determinants of performance in the traditional building process. *Construction Management and Economics*, **10**, 343–59.

Morris, P.W.G. (1972) *A Study of Selected Building Projects in the Context of Theories Organisation*. PhD Thesis, Manchester: University of Manchester, Institute of Science and Technology.

Morris, P.W.G. (1994) *The Management of Projects*. London: Thomas Telford.

Mortledge, R. (2006) Partnering – only good for a few? *RICS Construction Bulletin*, February.

Munns, A.K. (1995) Potential influence of trust on the successful completion of a project. *International Journal of Project Management*, **13**, 1.

Munns, A.K. (1996) Measuring Mutual Confidence in UK Construction Projects. *Journal of Management in Engineering*, **12**, 1.

Munns, A.K. & Bjeirmi, B.F. (1996) The role of project management in achieving project success. *International Journal of Project Management*, **14**, 2.

Munro, R. (2003) Disorganisation. In: Westwood, R. & Clegg, S. (eds) *Debating Organisations: Point-Counterpoint in Organisation Studies*. Oxford: Blackwell.

Murnighan, J.K., Malhotra, D. & Weber, M. (2004) Paradoxes of trust: Empirical and theoretical departures from a traditional model. In: Kramer, R. & Cook, K. (eds) *Trust and Distrust in Organisations*. New York: Russell Sage Foundation.

Myers, D. (2005) A review of construction companies' attitude to sustainability. *Construction Management and Economics*, **23**, 781–5.

Nahapiet, H. & Nahapiet, J. (1985) A comparison of contractual arrangements for building projects. *Construction Management and Economics*, **3**, 217–31.

Naoum, S. (2001) *People and organizational management in construction*. London: Thomas Telford.

Napier, J.A. (1970) *A Systems Approach to the Swedish Building Industry*. Stockholm: The National Swedish Institute for Building.

National Economic Development Council (1964) *The Construction Industry*. London: HMSO.

National Economic Development Office (1975) *The Public Client and the Construction Industries*. London: HMSO.

National Economic Development Office (1976) *The Professions in the Construction Industries.* London: HMSO.

National Economic Development Office (1978) *Construction for Industrial Recovery.* London: HMSO.

National Economic Development Office (1983) *Faster Building for Industry.* London: HMSO.

National Economic Development Office (1985) *Thinking about Building.* London: HMSO.

National Economic Development Office (1987) *Faster Building for Commerce.* London: HMSO.

Naylor, J., Pritchard, R.D. & Ilgen, D.R. (1980) *A Theory of Behaviour in Organisations.* New York: Academic Press.

Neilsen, E. (1986) Empowerment strategies: balancing authority and responsibility. In: Sirwastra, S. (ed.) *Executive Power.* San Francisco: Jossey-Bass.

Newcombe, R. (1994) Procurement paths – a power paradigm. In: Rowlinson, S.M. (ed.) *Proceedings of CIB W92 Symposium: East Meets West: Procurement Systems.* CIB Publication No. 175. Hong Kong: Department of Surveying, University of Hong Kong.

Newcombe, R. (1996) Empowering the construction project team. *International Journal of Project Management,* **14,** 2.

Newcombe, R. (1997) Procurement Paths – A cultural/political perspective. In: Davidson, C.H. & Meguid, T.A. (eds). *Procurement – a Key to Innovation.* Proceedings of CIB W92, Montreal: IF Research Corporation.

Newcombe, R. (2003) From client to project stakeholders: a stakeholder mapping approach. *Construction Management and Economics,* **21,** 841–8.

Nicolini, D., Holti, R. & Smalley, M. (2001) Integrating project activities: the theory and pratice of managing the supply chain through clusters. *Construction Management and Economics,* **19,** 37–47.

O'Brien, W.J., Fischer, M.A. & Jucker, J.V. (1995) An economic view of project coordination. *Construction Management and Economics,* **13,** 393–400.

Ogbonna, E. & Harris, L.C. (1998) Managing Organisational Culture: Compliance or Genuine Change? *British Journal of Management,* **9,** 273–88.

Oliver, C. (1997) The influence of institutional and task environment relationships on organizational performance: The Canadian construction industry. *Journal of Management Studies,* **34,** 1.

Olmstead, J.A. (2002) *Creating the Functionally Competent Organisation: An Open Systems Approach.* Westport, CT: Quorum Books.

Othman, A., Hassan T. & Pasquier, C. (2004) Drivers for dynamic brief development in construction. *Engineering, Construction and Architectural Management,* **11,** 4.

Palmer, I. & Hardy, C. (2000) *Thinking about Management: Implications of Organizational Debates for Practice.* London: Sage.

Partington, D., Pellegrinelli, S. & Young, M. (2005) Attributes and levels of programme management competence: an interpretive study. *Engineering, Construction and Architectural Management,* **23,** 2.

Peter, L.J. & Hull, R. (1969) *The Peter Principle.* New York: Morrow.

Pfeffer, J. (1978) The micro politics of organisations. In: Meyer, M.W. (ed.) *Environments and Organisations.* San Francisco: Jossey-Bass.

Pfeffer, J. (1981) *Power in Organisations.* Marshfield, MA: Pitman.

Pfeffer, J. (1981) Management as symbolic action: The creation and maintenance of organisational paradigms. In: Cummings, M. & Staw, B. (eds) *Research in Organisational Behaviour,* Vol. 3. Greenwich, CT: JAI Press.

Pfeffer, J. (1992) *Managing with Power: Politics and Influence in Organisations.* Boston, MA: Harvard Business School Press.

Pfeffer, J. & Salanick, G.R. (1978) *The External Control of Organisations.* New York: Harper & Row.

Phillips, R. & Lupton, S. (2000) *The Architect's Plan of Work*. London: RIBA Publications.

Pinto, J.K. (2000) Understanding the role of politics in successful project management. *International Journal of Project Management*, 18, 85–91.

Pocock, J.B., Liu, L.Y. & Tang, W.H. (1997) Prediction of project performance based on degree of interaction. *Journal of Management in Engineering*, 13, 2.

Poirot, J.W. (1991) Organising for quality: matrix organisation. *Journal of Management in Engineering*, 7, 178–86.

Porter, L.W. & Lawler, E.E. (1968) *Managerial Attitudes and Performance*. Homewood, IL: Dorsey.

Pries, F. & Janszen, F. (1995) Innovation in the construction industry: the dominant role of the environment. *Construction Management and Economics*, 13, 43–51.

Pryke, S. (2004) Analysing construction project coalitions: exploring the application of social network analysis. *Construction Management and Economics*, 22, 787–97.

Pryke, S. (2005) Towards a social network theory of project governance. *Construction Management and Economics*, 23, 927–39.

Quinn, R.E. (1988) *Beyond Rational Management: Mastering the Paradoxes and Competing Demands of High Performance*. San Francisco: Jossey-Bass.

Rahman, M. & Kumaraswamy, M. (2002) Joint risk management through transactionally efficient relational contracting. *Construction Management and Economics*, 20, 45–54.

Reve, T. & Levitt, R.E. (1984) Organisation and governance in construction. *International Journal of Project Management*, 2, 17–25.

Ries, C.J. (1964) *The Management of Defense*. Baltimore: Johns Hopkins Press.

Robbins, S. (2005) *Essentials of Organisational Behaviour*. Upper Saddle River, NJ: Pearson.

Roberts, C.C. (1989) Intermittent project management. *Journal of Management in Engineering*, 5, 84–9.

Roberts, P.W. & Greenwood, R. (1997) Integrating transaction cost and institutional theories: Toward a constrained-efficiency framework for understanding organizational design adoption. *Academy of Management Review*, 22, 2.

Robins, M.J. (1993) Effective project management in a matrix-management environment. *International Journal of Project Management*, 11, 11–14.

Rosenhead, J. (1998) *Complexity Theory and Management Practice*. LSE Working Paper 98.25. London: London School of Economics.

Rouleau, L. & Sequin, F. (1995) Strategy and organisation theories: common forms of discourse. *Journal of Management Studies*, 32, 22–38.

Rowlinson, S. (2001) Matrix organisational structure, culture and commitment: a Hong Kong public sector case study of change. *Construction Management and Economics*, 19, 669–73.

Rowlinson, S.M. (1988) *An Analysis of Factors Affecting Project Performance in Industrial Building*. PhD Thesis, Uxbridge: Brunel University.

Rowlinson, S.M. & Procter, S. (1997) Efficiency and Power: Organisational Economics Meets Organisation Theory. *British Journal of Management*, 8, Special Issue, S31–S42.

Rowlinson, S.M., Ho, T.K.K. & Yuen, P.H. (1993) Leadership style of construction managers in Hong Kong. *Construction Management and Economics*, 11, 455–65.

Royal Institute of British Architects (1997) *Plan of Work for Design Team Operation*. London: RIBA Publications Ltd.

Royer, I. (2005) Why bad projects are so hard to kill. In: *Harvard Business Review on Managing Projects*. Boston, MA: Harvard Business School.

Ruegg-Sturm, J. (2005) *The New St. Gallen Management Model: Basic Categories of an Approach to Integrated Management*. Basingstoke: Palgrave Macmillan.

Schein, E.H. (1992) *Organisational Culture and Leadership*. San Francisco: Jossey-Bass.

Schein, E.H. (1996) Culture: the missing concept in organisation studies. *Administrative Science Quarterly*, **41**, 229–40.

Schermerhorn, J., Jr., Hunt, G. & Osborn, R. (2004) *Core Concepts of Organisational Behaviour*. Hoboken, NJ: Wiley.

Schriberg, A., Schriberg, D. & Kumari, R. (2005) *Practicing Leadership*. Hoboken, NJ: Wiley.

Scott, J. (1992) *Social Network Analysis: A Handbook*. London: Sage.

Scott, W.R. (1992) *Organisations: Rational, Natural and Open Systems*, 3rd edn. Englewood Cliffs, NJ: Prentice-Hall.

Scott, W.R. (1995) *Institutions and Organisations*. Thousand Oaks, CA: Sage.

Scott, W.R. (1998) *Organisations: Rational, Natural and Open Systems*, 4th edn. Englewood Cliffs, NJ: Prentice Hall.

Scott, W.R. (2003) *Organisation: Rational, Natural and Open Systems*, 5th edn. Upper Saddle River, NJ: Prentice Hall.

Seifers, L. von (1972) *A Contingency Theory Approach to Temporary Management Systems*. PhD Thesis, Cambridge, MA: Graduate School of Business Administration, Harvard University.

Shaw, R. (1997) *Trust in the Balance*. San Francisco, CA: Jossey-Bass.

Shiers, D., Rapson, D., Roberts, C. & Keeping, M. (2006) Sustainable construction: the development and evaluation of an environmental profiling system for construction products. *Construction Management and Economics,* **24**, 1177–84.

Shirazi, B., Langford, D.A. & Rowlinson, S.M. (1996) Organisational structures in construction industry. *Construction Management and Economics*, **14**, 199–212.

Shoura, M.M. & Singh, A. (1998) Motivation parameters for engineering managers using Maslow's Theory. *Journal of Management in Engineering*, **15**, 5.

Simon, H.A. (1947) *Administrative Behaviour*. New York: Free Press.

Slevin, D.P. (1983) Leadership and the project manager. In: Cleland, D.I. & King, W.R. (eds) *Project Management Handbook*. New York: Van Nostrand Reinhold.

Smith, J.M., Kenley, R. & Wyatt, R. (1998) Evaluating the client briefing problem: an exploratory study. *Engineering, Construction and Architectural Management*, **5**, 4.

Smith, K.G., Carroll, S.J. & Ashford, S.J. (1995) Intra and interorganisational cooperation: towards a research agenda. *Academy of Management Journal*, **38**, 7–23.

Soderlund, J. (2004) On the broadening scope of the research on projects: a review and a model for analysis. *International Journal of Project Management*, **22**, 8.

Soetanto, R. & Proverbs, D. (2002) Modelling the satisfaction of contractors: the impact of client performance. *Engineering, Construction and Architectural Management*, **9**, 5/6.

Sorge, A. (1991) Strategic fit and the societal effect: interpreting cross-national comparisons of technology, organisation and human resources. *Organisation Studies*, **12**, 2.

Stacey, R. (1993) *Managing the Unknowable: Strategic Boundaries Between Order and Chaos in Organisations*. San Francisco: Jossey-Bass.

Stacey, R. (1996) *Strategic Management and Organisational Dynamics*. London: Pitman Publishing.

Stacey, R., Griffin, D. & Shaw, P. (2000) *Complexity and Management*. London: Routledge.

Staw, B. & Ross, J. (2005) Knowing when to pull the plug. In: *Harvard Business Review on Managing Projects*. Boston, MA: Harvard Business School.

Stinchcombe, A. & Heimer, C. (1985) *Organisation Theory and Project Management*. Oslo: Norwegian University Press.

Strategic Forum for Construction (2002) *Accelerating Change*. London: Rethinking Construction.

Strategic Forum for Construction (2006) *Strategic Forum for Construction*. *www.strategicforum.org.uk*, accessed on 4th October, 2006.

Tannenbaum, R. & Schmidt, W.H. (1973) How to choose a leadership pattern. *Harvard Business Review*, **51**, 162–80.

Tavistock Institute (1966) *Interdependence and Uncertainty*. London: Tavistock Publications.

Taylor, F.W. (1911) *The Principles of Scientific Management*. New York: Harper.

Thompson, J. (1967) *Organisation in Action*. New York: McGraw-Hill.

Thompson, N. (1978) Alternative methods of management. *Building*, 27 January, 67–70.

Thompson, P. (1991) The client role in project management. *International Journal of Project Management*, **9**, 90–92.

Tolbert, P.S. & Zucker, L.G. (1983) Institutional sources of change in the formal structure of organisations: the diffusion of civil service reforms, 1880–1935. *Administrative Science Quarterly*, **28**, 22–39.

Townsend, R. (1984) *Further Up The Organisation*. London: Michael Joseph.

Turner, J. (2004) Farsighted project contract management: incomplete in its entirety. *Construction Management and Economics*, **22**, 75–83.

Vroom, V.H. (1964) *Work and Motivation*. Malabar, FL: Robert E. Krieger Publishing Co.

Walker, A. (1976) *Project Management: A Review of the State of the Art*. London: The Institute of Quantity Surveyors (now the Royal Institution of Chartered Surveyors).

Walker, A. (1980) *A Model of the Design of Project Management Structures for Building Clients*. PhD Thesis, Liverpool: Liverpool Polytechnic (now Liverpool John Moores University).

Walker, A. (1994) *Building the Future – The Story of the Controversial Construction of the Campus of the Hong Kong University of Science and Technology*. Hong Kong: Longman Asia Ltd.

Walker, A. (1995) *Hong Kong: The Contractors' Experience*. Hong Kong: Hong Kong University Press.

Walker, A. & Chau, K.W. (1999) The relationship between construction project management theory and transaction cost economics. *Engineering, Construction and Architectural Management*, **6**, 2.

Walker, A. & Hughes, W.P. (1984) Private industrial project management: A systems-based case study. *Construction Management and Economics*, **2**, 93–110.

Walker, A. & Hughes, W.P. (1986) A conventionally-managed project: a systems-based case study. *Construction Management and Economics*, **4**, 57–74.

Walker, A. & Hughes, W.P. (1987a) A project managed by a multidisciplinary practice: a systems-based case study. *Construction Management and Economics*, **5**, 123–40.

Walker, A. & Hughes, W.P. (1987b) An analysis of the management of a public sector project: a systems-based case study. In: Lansley, P. & Harlow, P. (eds) *Managing Construction Worldwide*. London: Spon.

Walker, A. & Kalinowski, M. (1994) An anatomy of a Hong Kong project – organisation, environment and leadership. *Construction Management and Economics*, **12**, 191–202.

Walker, A. & Newcombe, R. (2000) The positive use of power on a major construction project. *Construction Management and Economics*, **18**, 37–44.

Walker, A. & Wilson, A.J. (1983) An approach to the measurement of the performance of the project management process. In: Chiddick, D. & Millington, A. (eds) *Proceedings of the Land Management Research Conference*. London: Spon.

Walker, D.H.T. (1994) Procurement systems and construction time performance. In: Rowlinson, S.M. (ed.) *Proceedings of CIB W92 Symposium: East Meets West: Procurement Systems*. CIB Publication No. 175. Hong Kong: Department of Surveying, University of Hong Kong.

Ward, S.C., Curtis, B. & Chapman, C.B. (1991) Objectives and performance in construction projects. *Construction Management and Economics*, **9**, 343–53.

Weber, M. (1947 trans.) *The Theory of Social and Economic Organisation* (ed. A.H. Henderson & T. Parsons). Glencoe, IL: Free Press (first published in 1924).

Weber, M. (1968 trans.) *Economy and Society: An Interpretive Sociology*, 3 vols (eds. Guenther Roth & Claus Wittich). New York: Bedminster Press (first published in 1924).

Weick, K.E. (1969) *The Social Psychology of Organising.* Reading, MA: Addison-Wesley.

Westwood, R. & Clegg, S. (2003) *Debating Organisation: Point-Counterpoint in Organisational Studies.* Oxford: Blackwell.

Wilkinson, S. (2001) An analysis of the problems faced by project management companies managing construction projects. *Construction and Architectural Management,* **8**, 3.

Williams, T.M. (1997) Empowerment vs risk management? *International Journal of Project Management,* **15**, 4.

Williamson, O.E. (1975) *Markets and Hierarchies: Analysis and Antitrust Implications.* New York: Free Press.

Williamson, O.E. (1979) Transaction-cost economics: the governance of contractual relations. *The Journal of Law and Economics,* **22**, 233–61.

Williamson, O.E. (1981a) The economics of organisation and the transaction cost approach. *American Journal of Sociology,* **87**, 548–77.

Williamson, O.E. (1981b) The modern corporation: origins, attributes. *Journal of Economic Literature,* **19**, 1537–68.

Williamson, O.E. (1985) *The Economic Institutions of Capitalism.* New York: Free Press.

Williamson, O.E. (1990) Chester Barnard and the incipient science of organisation. In: Williamson, O.E. (1996) *The Mechanism of Governance.* New York: Oxford University Press.

Williamson, O.E. (1999) Strategy research: governance and competence perspectives. *Strategic Management Journal,* **20**(12).

Williamson, O.E. (2000) The new institutional economics: taking stock, looking ahead. *Journal of Economic Literature,* **38**(3).

Wilson, A.J. (1976) Thirteenth century project management. *Building Technology and Management,* **7**, 5–8.

Winch, G. (1989) The construction firm and the construction project: a transaction cost approach. *Construction Management and Economics,* **7**, 331–45.

Winch, G. (2001) Governing the project process: a conceptual framework. *Construction Management and Economics,* **19**, 799–808.

Winch, G. (2002) *Managing Construction Projects: An Information Processing Approach.* Oxford: Blackwell.

Winch, G. (2006) Towards a theory of construction as production by projects. *Building Research and Information,* **34**, 2.

Winch, G., Usmani, A. & Edkins, A. (1998) Towards total project quality: a gap approach. *Construction Management and Economics,* **16**, 193–207.

Wong, E., Then, D. & Skitmore, M. (2000) Antecedents of trust in intra-organisational relationships within three Singapore public sector construction project management agencies. *Construction Management and Economics,* **18**, 797–806.

Wong, P. & Cheung, S. (2004) Trust in construction partnering: views from parties of the partnering dance. *International Journal of Project Management,* **22**, 6.

Wood, G. & Ellis, R. (2005) Main contractor experiences of partnering relationships on UK construction projects. *Construction Management and Economics,* **23**, 317–25.

Wood, G. & McDermott, P. (1999) Searching for trust in the UK construction industry: An interim view. In: *Proceedings of CIB W92 International Procurement Systems Conference.*

World Commission on Environment and Development (1987) *Our Common Future.* New York: Oxford University Press.

Yates, D. (1998) *Conflict and dispute resolution in the Hong Kong construction industry: A transaction cost economics perspective* M.Phil. Thesis, Hong Kong: University of Hong Kong.

Yeo, K.T. (1993) Systems thinking and project management – time to reunite. *International Journal of Project Management,* **11**, 111–17.

Youker, R. (1992) Managing the international project environment. *International Journal of Project Management,* **10**, 219–26.

Index